I0502930

Water Quality of the Chokosna, Gilahina, Lakina Rivers, and Long Lake Watershed along McCarthy Road, Wrangell-St. Elias National Park and Preserve, Alaska, 2007–08

By Timothy P. Brabets, Robert T. Ourso, Matthew P. Miller, and Anne M. Brasher

Prepared in cooperation with the National Park Service

Scientific Investigations Report 2011-5185

U.S. Department of the Interior
U.S. Geological Survey

U.S. Department of the Interior
KEN SALAZAR, Secretary

U.S. Geological Survey
Marcia K. McNutt, Director

U.S. Geological Survey, Reston, Virginia, 2011

For more information on the USGS—the Federal source for science about the Earth, its natural and living resources, natural hazards, and the environment, visit http://www.usgs.gov or call 1–888–ASK–USGS.

For an overview of USGS information products, including maps, imagery, and publications, visit http://www.usgs.gov/pubprod

To order this and other USGS information products, visit http://store.usgs.gov

Suggested citation:
Brabets, T.P., Ourso, R.T., Miller, M.P., and Brasher, A.M., 2011, Water quality of the Chokosna, Gilahina, Lakina Rivers, and Long Lake watershed along McCarthy Road, Wrangell-St. Elias National Park and Preserve, Alaska, 2007–08: U.S. Geological Survey Scientific Investigations Report 2011-5185, 56 p.

Contents

Figures

Tables

Conversion Factors, Datums, and Abbreviations and Acronyms

Conversion Factors

Multiply	By	To obtain
Length		
inch (in.)	2.54	centimeter (cm)
inch (in.)	25.4	millimeter (mm)
foot (ft)	0.3048	meter (m)
mile (mi)	1.609	kilometer (km)
Area		
square mile (mi^2)	2.590	square kilometer (km^2)
Volume		
cubic foot (ft^3)	0.02832	cubic meter (m^3)
cubic yard (yd^3)	0.7646	cubic meter (m^3)
Flow rate		
foot per second (ft/s)	0.3048	meter per second (m/s)
cubic foot per second (ft^3/s)	0.02832	cubic meter per second (m^3/s)
cubic foot per second per square mile [(ft^3/s)/mi^2]	0.01093	cubic meter per second per square kilometer [(m^3/s)/km^2]

Temperature in degrees Celsius (°C) may be converted to degrees Fahrenheit (°F) as follows:

$$°F=(1.8×°C)+32.$$

Temperature in degrees Fahrenheit (°F) may be converted to degrees Celsius (°C) as follows:

$$°C= (°F-32)/1.8.$$

Specific conductance is given in microsiemens per centimeter at 25 degrees Celsius (µS/cm at 25 °C).

Concentrations of chemical constituents in water are given either in milligrams per liter (mg/L) or micrograms per liter (µg/L).

Datums

Vertical coordinate information is referenced to the North American Vertical Datum of 1988 (NAVD 88).

Horizontal coordinate information is referenced to the North American Datum of 1983 (NAD 83).

Altitude, as used in this report, refers to distance above the vertical datum.

Abbreviations and Acronyms

ADAS	Algal Data Analysis System
ADOT&PF	Alaska Department of Transportation and Public Facilities
ANILCA	Alaska National Interest Lands Conservation Act
CCME	Canadian Council of Ministers of the Environment
DOC	dissolved organic carbon
EPT	Ephemeroptera (mayfly), Plecotera (stonefly), and Trichoptera (caddisfly)
HCO_3	Bicarbonate
IDAS	Invertebrate Data Analysis System
Mn	manganese
N	nitrogen
NAWQA	National Water-Quality Assessment program
NH_3	ammonia
NH_4	ammonium
NMDS	non-metric multidimensional scaling
NSC	Nash-Sutcliffe coefficient
NTU	Nephelometric Turbidity Unit
NWQL	National Water-Quality Laboratory
P	phosphorus
PEC	probable effect concentration
PEL	probable effect level
QMH	Qualitative Multi-Habitat
RMSE	root mean square error
RTH	Richest Targeted Habitat
USEPA	U.S. Environmental Protection Agency
USGS	U.S. Geological Survey
WRST	Wrangell-St. Elias National Park and Preserve

Water Quality of the Chokosna, Gilahina, Lakina Rivers, and Long Lake Watershed along McCarthy Road, Wrangell-St. Elias National Park and Preserve, Alaska, 2007–08

By Timothy P. Brabets, Robert T. Ourso, Matthew P. Miller, and Anne M. Brasher

Abstract

The Chokosna, Gilahina, and Lakina River basins, and the Long Lake watershed are located along McCarthy Road in Wrangell–St. Elias National Park and Preserve. The rivers and lake support a large run of sockeye (red) salmon that is important to the commercial and recreational fisheries in the larger Copper River. To gain a better understanding of the water quality conditions of these watersheds, these basins were studied as part of a cooperative study with the National Park Service during the open water periods in 2007 and 2008.

Water type of the rivers and Long Lake is calcium bicarbonate with the exception of that in the Chokosna River, which is calcium bicarbonate sulfate water. Alkalinity concentrations ranged from 63 to 222 milligrams per liter, indicating a high buffering capacity in these waters. Analyses of streambed sediments indicated that concentrations of the trace elements arsenic, chromium, and nickel exceed levels that might be toxic to fish and other aquatic organisms. However, these concentrations reflect local geology rather than anthropogenic sources in this nearly pristine area.

Benthic macroinvertebrate qualitative multi-habitat and richest targeted habitat samples collected from six stream sites along McCarthy Road indicated a total of 125 taxa. Insects made up the largest percentage of macroinvertebrates, totaling 83 percent of the families found. Dipterans (flies and midges) accounted for 43 percent of all macroinvertebrates found. Analysis of the macroinvertebrate data by non-metric multidimensional scaling indicated differences between (1) sites at Long Lake and other stream sites along McCarthy Road, likely due to different basin characteristics, (2) the 2007 and 2008 data, probably from the higher rainfall in 2008, and (3) macroinvertebrate data collected in south-central Alaska, which represents a different climate zone. The richness, abundance, and community composition of periphytic algae taxa was variable between sampling sites. Taxa richness and diversity were highest at the Long Lake outflow site, suggesting that the lake may have contributed planktonic taxa to the periphytic community and (or) created physical and chemical conditions at the outlet that were favorable to a variety of taxa.

Long Lake is fed by groundwater and by clear water (non glacial) streams, resulting in relatively high Secchi-disc readings ranging from 17.5 to 23 feet. Depth profiles of water temperature in the lake show a strong stratification during the summer from the surface to about 13 feet, with temperatures ranging from 16 to 5 °C. Depth profiles of dissolved oxygen in the lake show a strong stratification between 26 and 33 feet, below which the concentrations of dissolved oxygen decrease from 10 to 2 milligrams per liter. Because the Long Lake outlet stream supports a large run of sockeye salmon and water temperature is an important factor during its life cycle, a logistic model was used to simulate 1998–2006 water temperatures at this site. Analysis of simulation results for 1998–2008 indicated no significant trends in water temperature. 2007 water temperatures were the highest during the 10-year period.

Introduction

Wrangell-St. Elias National Park and Preserve (WRST) is the largest unit of the National Park System (fig. 1). This park includes North America's largest assemblage of glaciers, as well as the largest collection of peaks with elevations higher than 16,000 ft. The landscape is dominated by parts of the Alaska, Wrangell, St. Elias, and Chugach Mountain ranges. WRST is one of four contiguous conservation units (including Kluane National Park Reserve in the Yukon Territory, Alsek-Tatshenshini Provincial Park in British Columbia, and Glacier Bay National Park and Preserve in southeast Alaska) spanning some 24 million acres that have been recognized by the United Nations as an International World Heritage Site. Several of North America's highest peaks are within the park and preserve, including Mt. St. Elias (18,008 ft) and Mt. Wrangell (14,163 ft), which is an active volcano. The Malaspina Glacier, located in the southeast part of WRST, is the longest piedmont lobe glacier in North America.

WRST is approximately 200 mi east of Anchorage and 120 mi northeast of Valdez (fig. 1). The park/preserve is bounded by the Gulf of Alaska on the south, Mentasta Mountains and the Tetlin National Wildlife Refuge on the north, the Canadian border on the east, and the Copper River on the west. Many visitors to WSRT travel to McCarthy, in the heart of the park to visit the historic Kennecott Mine. Road access to McCarthy is by the 61-mi long McCarthy Road; however, its gravel and dirt surface makes for slow travel and it usually takes 3 hours or more to travel one way from Chitina to McCarthy. Other hazards can make travel time even longer. For example, heavy rain can make the road muddy and slippery, potholes can form, sharp rocks can cause flat tires, and narrow and one-lane bridges make maneuvering large vehicles difficult.

The McCarthy and the Nabesna Roads (fig. 1) are the only public roads into WRST that provide reasonable access for park visitors and residents. The continued maintenance of these roads by the Alaska Department of Transportation and Public Facilities (ADOT&PF) is mutually beneficial to WRST, the State, local communities, and residents. The demands of increasing visitor use have prompted ongoing discussions concerning the improvement of McCarthy Road. WRST completed a Scenic Corridor Plan, incorporating the State of Alaska's plans for a major upgrade to the McCarthy Road. The plan called for the opening of scenic overlooks, the construction of pullouts and interpretive waysides, and the development of foot and bike trails.

Section 201 (a) of the 1980 Alaska National Interest Lands Conservation Act (ANILCA, 1980) states that WRST will be managed for the following purposes, among others: "To maintain the scenic beauty and quality of high mountain peaks, foothills, glacial systems, lakes and streams, valleys, and coastal landscapes in their natural state: to protect habitat for, and populations of fish and wildlife…" Fish in many parts of WRST are not abundant due to naturally harsh stream conditions such as high gradients, velocities, sediment loads, and winter ice conditions. However, the clear water streams that originate in WRST along the McCarthy Road serve an important function in perpetuating local fish populations (National Park Service, 1998), as well as fish populations of the Copper River. Lakes along the McCarthy Road contain Dolly Varden, sockeye salmon, coho salmon, grayling, lake trout, and burbot.

The McCarthy Road is the former Copper River and Northwestern Railway, which operated from 1911 to 1938. Improvements to the road would be considered a large construction project since the upgrades would need to meet ADOT&PF standards. The potential effects on water quality and aquatic habitat by construction activities, and the increase in road use that would likely result from those activities, pose several issues of concern, including:

Sedimentation: The effects on aquatic habitat from increased sedimentation are well understood. Sediment loading in a channel generally results in the downstream deposition of the sediment. Gravel substrates, a critical component for successful spawning, become clogged and eventually cemented in place. Such clogging also severely affects habitat for benthic invertebrates, an important food source for fish.

Flow regimes: Road construction on sloped terrain can significantly change the hydrology of a local stream. For example, roads often intercept subsurface flow on slopes and convert it to surface flow. Loss of base flow can be detrimental to fish. Icing or large floods can severely affect the geomorphology of the stream.

Fish passage: Culverts can act in a number of ways to significantly reduce or completely block fish access to an upstream watershed. Undersized culverts can result in excessive water velocities at each end and through the culvert barrel. Such excessive velocities may completely prevent the entrance of a fish into the culvert, or exhaust a fish that has partially progressed upstream. Culverts installed at greater than the natural channel slope may also create excessive water velocities, especially at higher flows. Other culvert characteristics also determine fish passage status; for example, the roughness of the culvert material, and the size of the corrugation on steel culverts has a direct effect on the velocity of water flowing inside the culvert.

The McCarthy Road crosses many small streams (drainage area less than 0.5 mi^2) and a few mid-size rivers (fig. 2). Many of these streams support sockeye salmon from the Copper River, a major commercial fishery. However, the hydrology of the streams and limnology of the lakes is essentially unknown. To gain a basic understanding of these characteristics in this part of WRST, the U.S. Geological Survey (USGS), in cooperation with WRST, conducted a water-quality investigation of a number of streams along the McCarthy Road and the Long Lake watershed from 2007 to 2008.

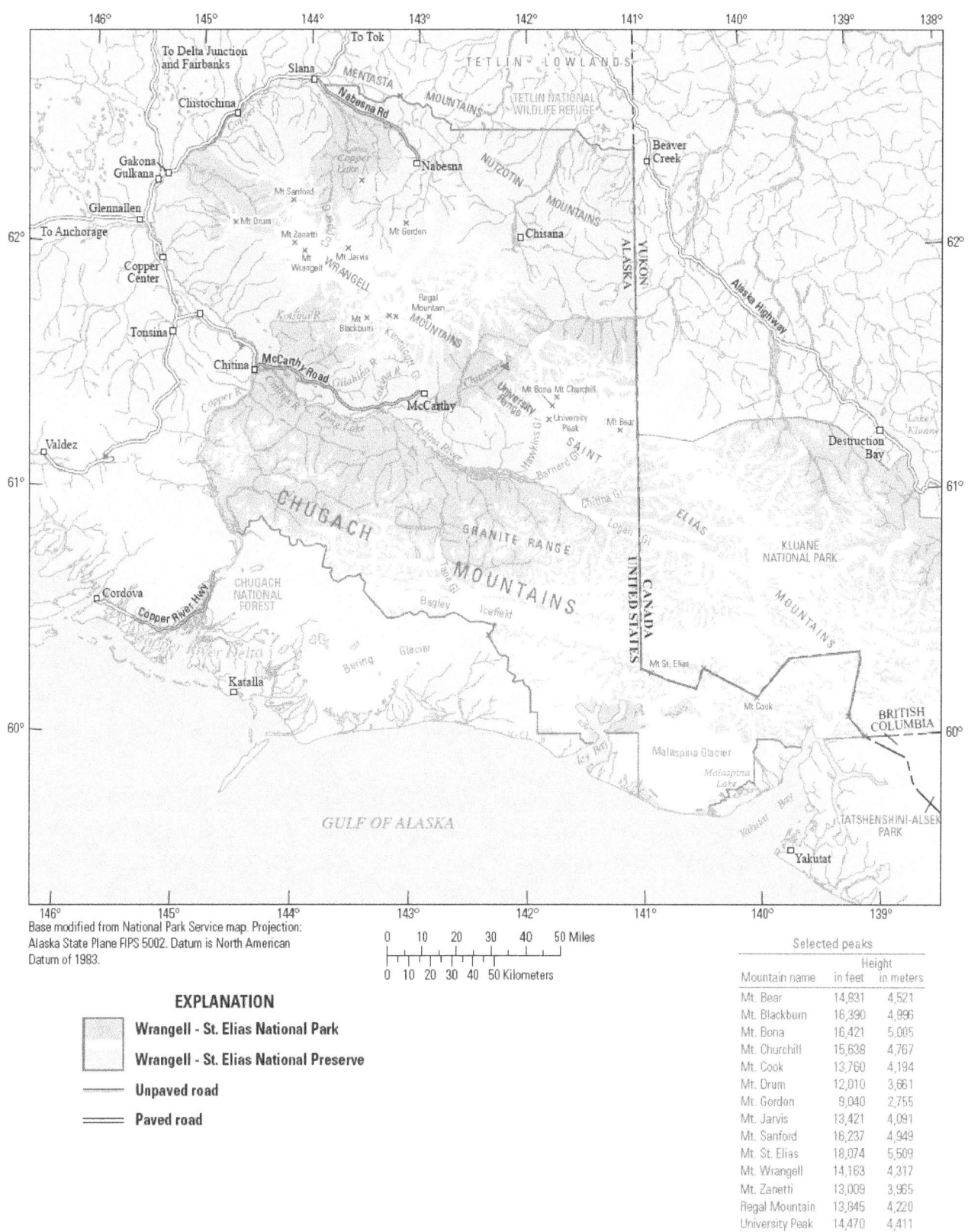

Figure 1. Location of Wrangell-St. Elias National Park and Preserve, Alaska.

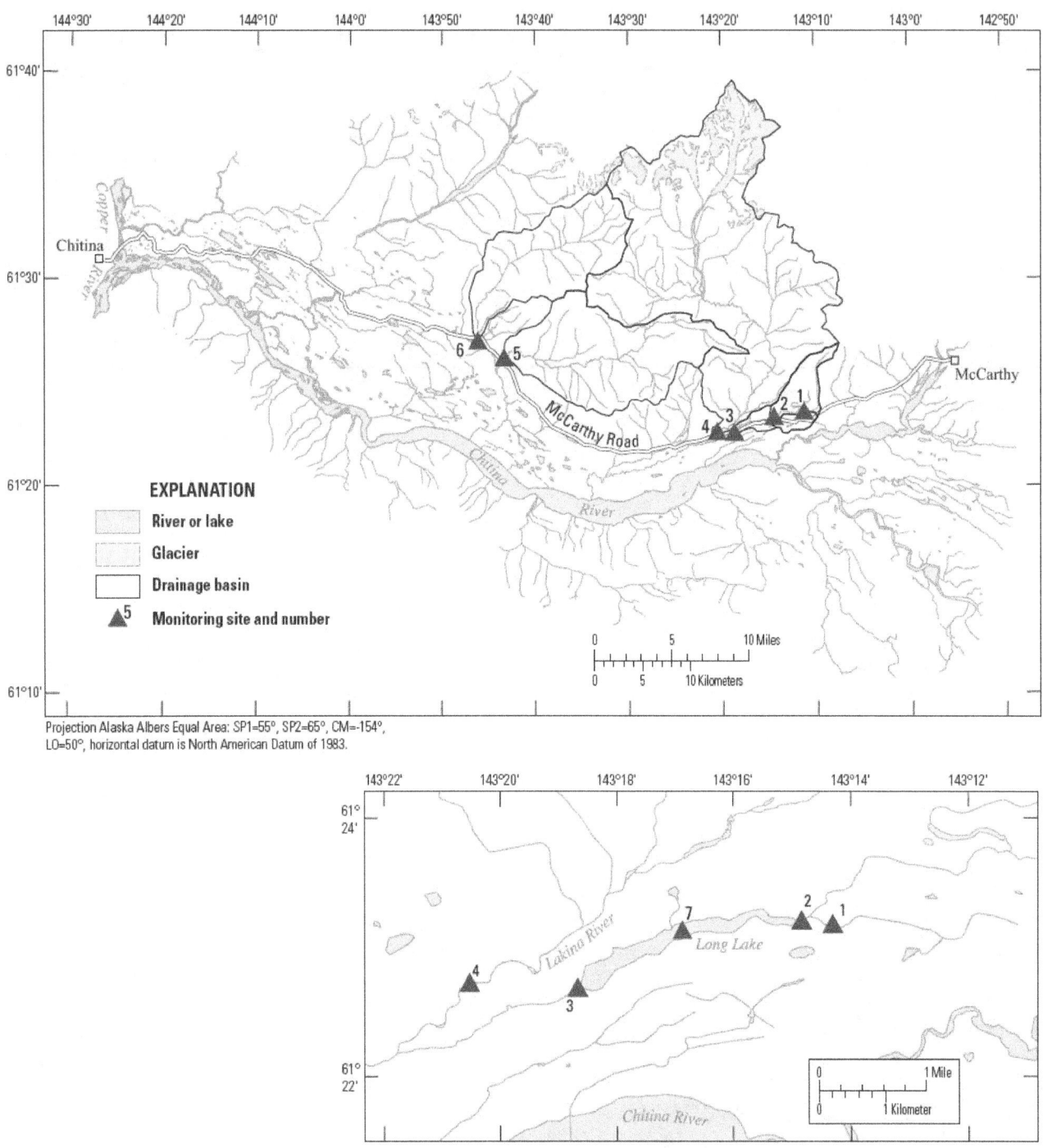

Projection Alaska Albers Equal Area: SP1=55°, SP2=65°, CM=-154°,
LO=50°, horizontal datum is North American Datum of 1983.

Figure 2. Locations of monitoring sites along McCarthy Road, Alaska.

Purpose and Scope

This report describes current water quality, physical habitat, and aquatic biology characteristics of selected streams and lakes along the McCarthy Road—focusing on the Chokosna, Gilahina, and Lakina Rivers, and Long Lake—that were collected as part of a joint WRST/USGS study in 2007 and 2008. This information will assist both WRST and ADOT&PF officials in planning improvements to the McCarthy Road in a manner that will minimize or avoid effects on the water quality and habitat of these streams and lakes.

Physical Setting

Because the Chugach Mountains act as a barrier, the climate of the study area is minimally affected by maritime influences. The climate is continental, with large temperature variability, low humidity, and relatively light and irregular precipitation. Summers are generally warm and sunny and winters are cold. Gulkana, a long-term representative National Weather Service index station for the study area and the upper Copper River Basin (fig. 1), has an average high temperature in July of 69 °F and an average low temperature in January of -13 °F. Average annual air temperature is 27 °F, average annual precipitation is 11 in., and average annual snowfall is 57 in. (Shulski and Wendler, 2007). Much of the region's annual precipitation falls in summer as sporadic, local rain showers or thunderstorms. In late summer, the rainfall occurs over larger areas since it originates from large low-pressure systems that move across the northern Pacific Ocean and onto the Alaska mainland, typically from the southwest. During the study period, total average rainfall from May through September, 2007, was 7.51 in., with July, August, and September rainfall totals of 1.17, 0.96, and 3.83 in., respectively. Total average rainfall in 2008 was 8.73 in. In contrast to 2007, July, August, and September rainfall totaled 3.50, 3.53, and 0.75 in., respectively.

The Chokosna, Gilahina, and Lakina River basins, and the Long Lake watershed (fig. 2) all drain the south side of the Wrangell Mountains. Although the watersheds are adjacent to each other, their basin characteristics are different. For example, the Chokosna and Gilahina River watersheds contain no glaciers; however, approximately 13 percent of the Lakina River watershed consists of glaciers. Glacier-fed rivers such as the Lakina River have different flow and water quality characteristics than non-glacier-fed streams.

The Chokosna River basin (39.2 mi^2) consists of surficial deposits of glacial and glaciolacustrine origin, and sedimentary and volcanic rocks of the Station Creek and McCarthy Formation. Adjacent to the Chokosna River basin, the Gilahina River basin (51.6 mi^2) consists of sedimentary and volcanic rocks of the Station Creek Formation to the north and intrusive rocks, primarily gabbro, to the south. The valley floor is covered by surficial deposits of glacial and glaciolacustrine origin. Adjacent to the Gilahina River basin is the Lakina River basin (141 mi^2), which is underlain by several types of geologic materials, including unconsolidated surficial deposits; volcaniclastic deposits from the Wrangell Volcanic Field; and sedimentary, volcanic, intrusive, and metamorphic rocks from what is termed (by Richter and others, 2006) the Wrangellia Terrane. The Long Lake watershed (11 mi^2) consists primarily of the surficial glacial and glaciolacustrine deposits with about 1 mi^2 of sedimentary and volcanic rocks from the Pleistocene and early and late Cretaceous age. The reader is referred to Richter and others (2006) for a more comprehensive view of the geology.

Long Lake, at approximately Mile 50 of the McCarthy Road, is about 2 mi long and about 650 ft wide (fig. 3). The deepest part of the lake is about 60 ft, near the outlet. A stream draining about 9.0 mi^2 from the east contributes most of the inflow to Long Lake and the remaining 2 mi^2 of the watershed drains from the north. Long Lake is partly fed by groundwater, and the watershed is dominated by wetlands. Substantial runs of sockeye salmon spawn in Long Lake (table 1).

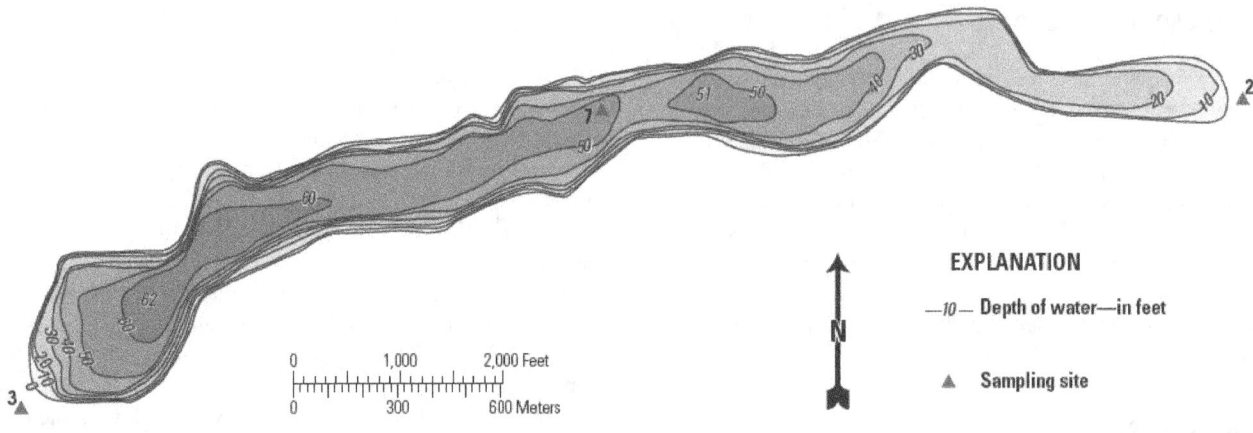

Figure 3.   Bathymetry of Long Lake near McCarthy, Alaska.

Table 1.   Annual number of salmon passing through the weir at Long Lake near McCarthy, Alaska, 1974-2008.

Year	Long Lake weir salmon count	Year	Long Lake weir salmon count
1974	4,684	1992	10,091
1975	6,768	1993	16,101
1976	24,689	1994	18,289
1977	8,624	1995	17,923
1978	15,458	1996	6,309
1979	46,110	1997	4,443
1980	39,038	1998	8,441
1981	12,659	1999	12,922
1982	28,047	2000	8,645
1983	28,133	2001	26,999
1984	10,637	2002	49,747
1985	21,131	2003	4,604
1986	16,997	2004	19,215
1987	13,633	2005	7,770
1988	7,543	2006	9,239
1989	14,981	2007	7,846
1990	21,664	2008	831
1991	11,511		

Methods of Data Collection and Analysis

A wide variety of water quality and biological data were collected during the open water periods of 2007 and 2008 from the selected river and lake sites (fig. 2, table 2). Water samples collected from the rivers and Long Lake were analyzed for major ions, total dissolved solids, total and dissolved nutrients, dissolved organic carbon, phytoplankton, chlorophyll-a, and trace elements (table 3). The field-collection and processing equipment used was made from Teflon, glass, or stainless steel to prevent sample contamination and to minimize analyte losses through adsorption. All sampling equipment was cleaned prior to use with a non-phosphate laboratory detergent and deionized water, followed by a native-water rinse prior to sample collection. Depth-integrated water samples were collected across the stream using the equal-width-increment method (U.S. Geological Survey National Field Manual, variously dated) and processed in the field using methods and equipment described by Shelton (1994). Samples were collected from Long Lake using a 3.0 liter acrylic Van Dorn sampler. Samples to be analyzed for dissolved constituents were filtered through 0.45-micrometer (μm) capsule filters. Water samples were sent to the USGS National Water-Quality Laboratory for analysis using standard USGS analytical methods (Fishman and Friedman, 1989; Patton and Truitt, 1992; Fishman, 1993).

A Yellow Springs Instrument multi-parameter sonde was used to measure water temperature, dissolved-oxygen concentration, specific conductance, turbidity, and pH at the time of sampling. Discharge measurements also were made at the time of sampling using methods of Turnipseed and Sauer (2010). Instruments were installed at all sites to collect water temperature on an hourly basis. At the time of sampling, the instruments would be cleaned and compared to the field measurement to insure accurate readings. Adjustments to the continuous water-temperature data were made using methods outlined by Wagner and others (2006).

The water-quality and temperature data were compared to known or published guidelines that are recommended for fish survival to determine if particular stream conditions may be affecting aquatic life. Water-quality data were analyzed to determine if seasonal patterns exist and to identify the basic composition of each particular stream. Water-temperature data were compared with air-temperature data from a climate station located at McCarthy and used in a logistic regression to determine a 10-year time series of water temperature.

Samples of streambed sediments were collected from several depositional areas at each site and analyzed for trace elements. The samples were collected from the surface of the streambed using Teflon tubes or Teflon coated spoons and composited in glass bowls (Shelton and Capel, 1994). Because the concentration of trace elements from streambed materials is strongly affected by the particle-size distribution of the sample, only that portion of the sample that was finer than 63 μm was analyzed. Stream water was used for sieving the trace-element sample through a 63-μm mesh. Water included in the trace-element sample was decanted after very fine-grained sediments had settled (about 24 hours). Trace elements in streambed sediments were analyzed following a total digestion procedure as described by Arbogast (1990). The trace-element concentrations in streambed sediments then were compared with those collected by the USGS, National Water-Quality Assessment (NAWQA) Program, and with guidelines established by the Canadian Council of Ministers of the Environment (CCME) (1999).

Samples for benthic macroinvertebrates were collected at six different sites in the study area (table 2). Richest Targeted Habitat (RTH) and Qualitative Multi-Habitat (QMH) samples were collected and processed using protocols established by the NAWQA Program (Cuffney and others, 1993) and were identified by EcoAnalysts, Inc. in Moscow, Idaho. RTH samples are semi-quantitative, providing taxa abundance and richness measures by sampling from a delineated area to a predetermined depth of bed sediment, carefully removing macroinvertebrates, and then immediately washing those macroinvertebrates downstream into a fine (425-μm mesh) net. Five sites in stream riffles within the reach were selected in advance for collection. QMH sampling is a measure of the presence or absence of macroinvertebrates obtained by collecting samples with a standard D-frame net (210-μm mesh net) from all possible microhabitats to give a broad account of organisms within the stream reach.

Five of the six sites were sampled in 2007, and four of the six sites were sampled in 2008 (table 2). Sites 1 (Long Lake tributary 2) and 4 (Lakina River) were sampled only in 2007, as they were considered too dissimilar to the rest of the sites. Site 1 was located upstream of McCarthy Road as it issued from a very steep incline and then flowed into a roadside ditch before being channeled into a culvert under the road. After this site had been initially sampled (QMH techniques only), the discovery of the major inflow tributary to Long Lake led to the establishment of site 2 the following year. This site was more representative of the natural streams found throughout the area. Site 4, the Lakina River, was sampled in 2007 using only QMH techniques, as riffle areas were unavailable for RTH sampling. This site was not sampled in 2008 because the flow and velocity were too high for safe sampling. Because the Lakina River is a glacial river, it exhibits depauperate macroinvertebrate communities due to glacial fines and the subsequent deleterious effect on the growth of periphyton, a primary food source for many benthic macroinvertebrates, while also adversely affecting filtration by macroinvertebrates (Gislason and others, 2000). The Gilahina River was unable to be sampled using RTH techniques in 2008 as the high water level obscured any riffle areas and the Chokosna River was likewise too high to facilitate RTH sampling techniques in 2007. Long Lake Creek, site 3, was initially sampled using only RTH techniques on July 29, 2007, but was followed 2 days later by RTH and QMH sampling.

The macroinvertebrate data gathered at the McCarthy Road sites were used in the calculation of metrics for comparison against all other sites in Alaska similarly sampled using the NAWQA protocols. Converting both RTH and QMH data to metrics used for comparing or contrasting sites was accomplished using the Invertebrate Data Analysis System (IDAS) (Cuffney and Brightbill, 2011). The IDAS software is designed to parse, organize, and calculate metrics related to macroinvertebrate data collected as part of the NAWQA protocols. The software adds accuracy and precision to data interpretation by tracking and documenting settings used in metric generation, thereby yielding easily reproducible and comparable results. The options for data processing were set for the default, with the exception of the combination of parent taxa with children.

Macroinvertebrate data were further analyzed using the ordination technique, non-metric multidimensional scaling (NMDS) (Venables and Ripley, 2002; Oksanen and others, 2010) using the R software (R Development Core Team, 2010) and specifically the vegan (Oksanen and others, 2010) and MASS (Venables and Ripley, 2002) packages. Macroinvertebrate presence/absence data were standardized to the lowest identifiable taxonomic level for each sample in order to generate a Bray-Curtis dissimilarity (distance) input matrix. NMDS plots were then generated from the two distance matrices for the McCarthy Road samples, and the McCarthy Road samples combined with other Alaska macroinvertebrate samples.

QMH and RTH algae samples were collected according to protocols outlined by Porter and others (1993). The algal QMH sample is similar to the macroinvertebrate QMH sample; its purpose is to identify the species of algae present at multiple habitats within each stream reach. Algae samples were collected in habitats similar to the QMH and RTH macroinvertebrate samples (depositional zones, woody debris, and riffles) and were analyzed by the Philadelphia Academy of Natural Science. Periphytic algae QMH and RTH samples were collected in 2007 and 2008 from the inlet and outlet of Long Lake, as well as from the Lakina, Gilahina, and Chokosna Rivers (table 2). Preparation of the QMH and

RTH data was accomplished using the Algal Data Analysis System (ADAS) (Cuffney and Brightbill, USGS, written commun., 2010). Similar to IDAS, ADAS was designed to parse, organize, and calculate metrics related to algal data. Samples were kept separate while resolving ambiguities and parents were distributed among children for QMH and RTH data. Taxa richness for each sample was calculated using QMH and RTH data; abundance, diversity values, and multivariate analyses were generated using only the RTH data. Abundance data are reported for both density (number of individuals) and biovolume (algal biomass). The advantage of reporting abundance as biovolume in addition to density is that biovolume accounts for cell-size variation between taxa.

A similarity profile (SIMPROF) test (Clarke and others, 2008) was used to define groups of sites that are not statistically different from one another. The similarity in periphytic algal community structure between sites was investigated by generating a NMDS ordination plot based on a Bray-Curtis similarity resemblance matrix calculated using fourth-root transformed RTH biovolume data. Sites that are more similar to one another plot closer together and sites that are less similar to one another plot further apart. The PRIMER package (Clarke and Gorley, 2006) was used to generate the resemblance matrix and NMDS plot. The ability of the NMDS

to represent differences in community structure was assessed through calculation of a two-dimensional stress value. Stress values less than 0.05 are indicative of an excellent representation of the site relations, whereas values between 0.05 and 0.1 indicate a good representation, and values greater than 0.3 suggest that the locations of the points in the plot are random (Carke and Warwick, 2001).

Zooplankton samples were collected from Long Lake in May, June, August, and September 2008. Samples were collected using a 1.5-ft diameter, 153-μm mesh, conical plankton net. Vertical tows were made from 3 ft above the lake bottom to the surface and the contents were preserved in 10 percent buffered formalin. Preserved zooplankton samples were concentrated by gently pouring the original sample (about 125–200 mL) through a 153-μm mesh, followed by re-suspension into about 65 mL 4 percent formalin (final concentration). Following methods used by Balcer and others (1984), and Ward and others (1959), 1 to 2 mL subsamples were enumerated by low power microscopy (dissecting microscope) to obtain approximately 100 total animals in each subsample. Broad groupings of zooplankton such as Daphnia, Bosmina, and Ceriodaphnia were enumerated. Net diameter, depth and number of tows were factored in when calculating number of animals per cubic meter of water sampled.

Table 2. Water-quality and biology sample collection sites along McCarthy Road, Alaska.

[**Site No :** See figure 2 for location. **Abbreviations:** USGS, U.S. Geological Survey; mi^2, square mile; <, less than]

Site No.	USGS station No.	Latitude/longitude	Station name	Area (mi^2)	Glaciers (mi^2)	Lakes (mi^2)	Remarks
			Surface water sites				
1	15210250	61° 23' 14"143° 14' 23"	Long Lake Tributary 2 at McCarthy Road near McCarthy	2.0	0	0	Sampled in 2007
2	15210260	61° 23' 14" 143° 14' 46"	Long Lake Tributary 1, 150 ft above mouth near McCarthy	9.0	0	0	Sampled in 2008
3	15210300	61° 22' 43" 143° 18' 37"	Long Lake Creek at McCarthy Road near McCarthy	11.0	0	0.23	Sampled in 2007 and 2008
4	15210200	61° 22' 30" 143° 20' 50"	Lakina River near McCarthy	141.0	12	<1	Sampled in 2007
5	15210600	61° 26' 19" 143° 43' 00"	Gilahina River at McCarthy Road near Chitina	51.6	0	0	Sampled in 2007 and 2008
6	15210700	61° 27' 21" 143° 45' 43"	Chokosna River at McCarthy Road near Chitina	39.2	0	0	Sampled in 2007 and 2008
			Lake sites				
7		61° 23' 10" 143° 16' 45"	Long Lake site 1 near McCarthy				Sampled in 2008

Table 3. Analyses made on water samples and streambed sediments collected along McCarthy Road, Alaska, 2007 and 2008.

[**Abbreviations:** C, degrees Celsius; μS/cm, microsiemens per centimeter; mg/L, milligram per liter; NTU, Naphelometric Turbidity Unit; ft³/s, cubic foot per second; μg/L, microgram per liter; μg/g, microgram per gram]

Field measurements of water	Reporting level
Water temperature (°C)	0.1
Specific conductance (μS/cm)	5
pH	0.1
Dissolved oxygen (mg/L)	0.1
Turbidity (NTU)	1
Streamflow (ft³/s)	0.01

Chemical constituents in water	Reporting level (mg/L)
Alkalinity	1
Bicarbonate	1
Calcium	0.022
Chloride	0.06
Dissolved solids	12
Fluoride	0.04
Magnesium	0.008
Potassium	0.022
Silica	0.029
Sodium	0.06
Sulfate	0.09

Nutrients in water	Reporting level (mg/L)
Nitrogen, ammonia - dissolved	0.010
Nitrogen, ammonia + organic - dissolved and total	0.050
Nitrogen, NO_2 - dissolved	0.001
Nitrogen, $NO_2 + NO_3$ - dissolved	0.008
Phosphorus - dissolved	0.003
Phosphorus - total	0.004
Orthophosphate - dissolved	0.004

Trace elements in water	Reporting level (μg/L)
Iron	3.2
Manganese	0.16

Suspended sediment in water	Reporting level (mg/L)
Concentration	1

Organics in water	Reporting level
Dissolved organic carbon (mg/L)	0.15
Pheophytin phytoplankton (μg/L)	0.1
Chlorophyll-*a* phytoplankton (μg/L)	0.1

Trace elements in streambed sediments	Reporting level (μg/g)
Aluminum	850
Antimony	0.04
Arsenic	1
Barium	0.2
Beryllium	0.03
Bismuth	0.06
Cadmium	0.007
Cerium	0.1
Cesium	0.05
Chromium	0.5
Cobalt	0.02
Copper	2
Gallium	0.02
Iron	5
Lanthanum	0.05
Lead	0.4
Lithium	0.3
Manganese	0.7
Mercury	0.01
Molybdenum	0.05
Nickel	0.3
Niobium	1
Scandium	0.04
Selenium	0.1
Silver	2
Strontium	0.8
Thallium	0.08
Thorium	0.1
Titanium	40
Uranium	0.02
Vanadium	0.2
Yttrium	0.05
Zinc	3

Chemical constituents in streambed sediments	Reporting level
Inorganic carbon (percent)	0.01
Organic carbon (percent)	0.01
Total carbon (percent)	0.01
Calcium (μg/g)	100
Magnesium (μg/g)	6
Phosphorus (μg/g)	5
Potassium (μg/g)	20
Sodium (μg/g)	20
Sulfur (percent)	0.05

Water Quality Characteristics

Water-quality data collected at streams located along the McCarthy Road and at Long Lake in 2007 and 2008 included (1) field measurements of streamflow, specific conductance, pH, water temperature, turbidity, and dissolved-oxygen concentration, (2) collection and analysis of water samples for suspended sediment, major inorganic ions, nutrients, dissolved organic carbon (DOC), and chlorophyll-*a*, (3) collection and analysis of hourly water temperature data, and (4) collection and analysis of streambed sediments for trace elements.

Streamflow, Suspended Sediment, and Turbidity

Typical of most rivers in interior Alaska, most of the annual flow of the Chokosna, Gilahina, and Lakina Rivers occurs from mid-May to mid-October. Above average flows in June are dominated by snowmelt, and in July through September, by rainfall. The highest flows, ranging from 259 to 932 ft^3/s, were measured at the glacier-fed Lakina River (table 4). The lowest flows, ranging from 0.2 to 17 ft^3/s, were measured at the Long Lake inlet and outlet streams. Measured flows at the Gilahina and Chokosna Rivers were similar, with flows at the Gilahina River (the larger basin) ranging from 53 to 124 ft^3/s and at the Chokosna River ranging from 38 to 179 ft^3/s. During the study period, discharges measured in May 2007 were higher than discharges measured in May 2008, reflecting an early snowmelt period in 2007. Based on discharge per unit area (mi^2), runoff is quite similar among the Lakina, Gilahina, and Chokosna Rivers, ranging from 1.0 to 6.6 ft^3/s/mi^2.

Sediment in rivers is transported in the water column and along the bed (bedload). Suspended sediment generally consists of fine particles such as clay, silt, and fine sand that are transported in the stream while being held in suspension by the turbulence of flowing water. Bedload consists of coarse sediment particles such as sands, gravels, and sometimes cobbles that are transported along or near the streambed. Measured suspended sediment concentrations in the streams were low—less than 6 milligrams per liter (mg/L)—at the Long Lake inlet and outlet sites (table 4). Minimum concentrations of 1.0 mg/L were measured several times at these two sites. In turn, Secchi-disc transparencies, an indicator of the clarity of lake water, were relatively high, and ranged between 17.5 and 23 ft at Long Lake (table 5).

The highest measured concentrations of suspended sediment were 101 mg/L at the Lakina River, and 69 mg/L at the Chokosna River. Both of these measurements were made during the snowmelt period in May 2007, though the glacier fed Lakina River would be expected to have higher suspended sediment concentrations throughout the summer. The highest concentration (17 mg/L) at the Gilahina River also was noted at this time. The largest variation in suspended sediment concentrations occurred at the Chokosna River. Turbidity values ranged from less than 2 to 42 NTU and showed some correlation with the suspended sediment concentrations.

Water Temperature

Water temperature determines the amount of oxygen water can contain when at equilibrium with the atmosphere, and it also controls the metabolic and growth rates of fish. Sockeye salmon have adapted to specific spawning times and water temperatures so that incubation and emergence occur at the most favorable time of the year during spring and early summer (Reiser and Bjornn, 1979). For sockeye salmon, the range in water temperature for all life stages (migration, spawning, incubation, and rearing) ranges from 2.0 to 17.8 °C (Reiser and Bjornn, 1979).

Water temperatures of the Chokosna and Gilahina Rivers were similar. During the 2 years of data collection, recorded average daily water temperatures in the streams did not exceed 10 °C (table 4, fig. 4), which likely reflects the effect of snowmelt in the upper portion of these two basins. Water temperature at these two rivers also correlates with air temperature (fig. 4). Air temperature in 2007 was warmer than in 2008, and the Gilahina River water temperature between mid-July and mid-August was approximately 3 °C warmer in 2007. At the glacier-fed Lakina River, water temperatures were less than 9 °C (table 4, fig. 5), and there is some correlation with air temperature. However, the variability between 2007 and 2008 is not as large compared with the Gilahina and Chokosna Rivers. The most pronounced differences in water temperature were noted in the inlet and outlet streams of Long Lake (table 4, fig. 6). Water temperatures were measured as high as 15 °C in the inlet stream and as high as 19 °C in the outlet stream. In the Long Lake watershed, there are no perennial snowfields, and water temperatures correlate more strongly with air temperature.

Depth profiles of the water temperature of Long Lake indicate that temperatures change with depth throughout the summer months (table 5, fig. 7). In May 2008, just after ice out, water temperatures in the top 6 ft were greater than 5 °C, but quickly decreased with increasing depth to 4 °C. By July and August, water temperatures from the surface to about 13 ft deep were about 15 °C; a strong stratification was present between 13 and 20 ft, and water temperatures steadily decreased with depth to about 5 °C. By mid to late September, temperatures from the surface to 26 ft had cooled to about 10 °C, a weak stratification was present between 26 and 33 ft, and then water temperatures steadily decreased to 6 °C.

Table 4. Physical field parameters and suspended sediment measured at stream sites along McCarthy Road, Alaska, 2007 and 2008.

[**Abbreviations:** ft^3/s, cubic foot per second; (ft^3/s)/mi^2, cubic foot per second per square mile; mg/L, milligram per liter; µS/cm at 25 C, microsiemens per centimeter at 25 degrees Celsius; C, degrees Celsius; mi^2, square mile; NTU, Nephelometric Turbidity Unit; <, less than; –, no data]

Date	Time	Discharge (ft^3/s)	Discharge per unit area [(ft^3/s)/mi^2]	Dissolved oxygen (mg/L)	pH (units)	Specific conductance (µS/cm at 25°C)	Water temperature (°C)	Turbidity (NTU)	Suspended sediment (mg/L)
				Lakina River (drainage area 141 mi^2)					
05-22-07	1530	259	1.8	11.1	8.2	278	6.7	42	101
07-31-07	1000	932	6.6	–	–	–	–	–	–
10-02-07	1715	294	2.1	11.8	8.2	358	4.4	5	63
				Long Lake Inflow (drainage area 9.0 mi^2)					
05-22-07	1000	3.8	0.4	9.6	7.8	192	7.2	<2	1
07-30-07	0945	.2	.0	6.1	7.8	418	13.6	3.3	–
05-21-08	1030	2.1	.2	9.9	7.8	277	4.7	<2	<1
07-01-08	1000	1.9	.2	9.3	8.0	328	8.1	<2	1
08-06-08	1406	14	.2	9.8	7.8	267	9.2	3.7	4
09-24-08	1020	3.6	.4	10.8	7.8	338	5.2	<2	3
				Long Lake Outflow (drainage area 11.0 mi^2)					
05-22-07	1210	15	1.4	11.5	8.1	271	10.0	<2	1
07-30-07	1900	4	.4	12.3	8.6	334	20.4	<2	–
05-20-08	1730	8	.7	–	8.1	350	8.0	<2	1
07-01-08	1125	5.7	.5	10.9	8.5	371	15.6	<2	6
08-05-08	1644	17	1.5	10.4	8.3	361	14.6	<2	1
09-23-08	1745	11	1.0	11.4	8.3	362	10.6	<2	4
				Gilahina River (drainage area 51.6 mi^2)					
05-23-07	0900	96	1.9	12.9	8.1	173	3.1	6.4	17
08-01-07	1200	53	1.0	12.2	8.3	262	7.7	1.2	–
10-03-07	1600	88	1.7	13.1	8.2	248	2.8	–	5
05-19-08	1810	69	1.3	12.1	8.3	199	6.9	5.1	11
06-30-08	1730	124	2.4	11.5	8.3	224	8.8	2.8	11
09-24-08	1600	91	1.8	12.8	8.3	275	4.5	3.3	4
				Chokosna River (drainage area 39.2 mi^2)					
05-23-07	1100	85	2.2	13.2	8.1	236	3.1	37	69
07-29-07	1400	38	1.0	–	8.3	504	9.1	–	–
10-03-07	1240	66	1.7	13.1	8.2	513	1.9	–	11
05-19-08	1545	46	1.2	12.1	8.1	327	7.0	13	22
06-30-08	1540	90	2.3	11.6	8.2	441	8.6	14	34
08-04-08	1635	179	4.6	11.7	8.3	456	6.7	21	57
09-24-08	1700	89	2.3	12.6	8.3	522	4.4	–	4

Table 5.   Physical field parameters measured at Long Lake near McCarthy, Alaska, 2008.

[**Abbreviations:** ft, foot; mg/L, milligram per liter; μS/cm at 25 C, microsiemens per centimeter at 25 degrees Celsius; C, degree Celsius; NTU, Nephelometric Turbidity Unit; –, no data]

Date	Sample depth (ft)	Dissolved oxygen (mg/L)	Oxygen saturation (percent)	pH (units)	Specific conductance (μS/cm at 25°C)	Water temperature (°C)	Turbidity (NTU)	Chlorophyll-α	Secchi disk depth (ft)
05-20-2008	0.0	10.2	94	7.7	326	6.9	<2	–	
	3.3	10.2	93	7.7	326	6.5	<2	–	
	4.9	10.2	93	7.7	326	6.3	<2	–	
	6.6	8.9	84	7.7	344	4.8	<2	–	
	8.2	7.7	81	7.8	353	4.7	<2	–	
	9.8	6.8	81	7.7	364	4.5	<2	–	
	11.4	5.7	79	7.7	376	4.2	<2	–	
	13.1	5.1	78	7.7	384	4.0	<2	–	
	16.4	4.2	73	7.7	394	3.7	<2	–	
	19.7	3.9	70	7.7	395	3.7	<2	–	
	23.0	3.7	65	7.7	395	3.7	<2	–	
	26.2	3.5	63	7.7	396	3.6	<2	–	
	29.5	3.5	62	7.7	396	3.6	<2	–	
	32.8	3.4	62	7.7	397	3.6	<2	–	
	36.1	3.3	59	7.7	397	3.6	<2	–	
	39.4	3.0	56	7.7	397	3.6	<2	–	
	42.6	2.2	41	7.7	398	3.5	<2	–	
	45.9	1.7	31	7.6	400	3.4	<2	–	
	49.2	1.6	30	7.6	400	3.4	<2	–	
07-01-2008	0.0	9.9	106	8.4	372	16.0	<2	1.0	17.5
	1.6	9.9	106	8.4	372	16.0	<2	1.4	
	3.3	9.9	106	8.4	372	15.9	<2	1.1	
	4.9	10.0	106	8.4	372	15.8	<2	1.0	
	6.6	10.0	106	8.4	372	15.7	<2	1.3	
	8.2	9.9	106	8.4	372	15.7	<2	.9	
	9.8	10.0	106	8.4	372	15.6	<2	1.4	
	11.4	9.9	105	8.4	372	15.5	<2	1.0	
	13.1	10.0	105	8.4	372	15.5	<2	1.2	
	14.8	10.8	111	8.3	375	14.3	<2	1.2	
	16.4	12.2	117	8.3	375	11.2	<2	2.3	
	18.0	12.1	114	8.3	377	10.4	<2	2.2	
	19.7	12.1	111	8.2	379	9.4	<2	3.6	
	21.3	11.7	105	8.2	382	8.5	<2	5.7	
	23.0	10.9	97	8.1	383	8.1	<2	8.4	
	24.6	9.9	88	8.1	385	7.6	<2	9.4	
	26.2	9.2	80	8.0	385	7.3	<2	11.0	
	27.9	8.6	74	8.0	387	7.0	<2	13.3	
	29.5	8.1	70	8.0	387	6.8	<2	16.7	
	31.2	7.4	64	7.9	387	6.7	<2	15.0	
	32.8	7.0	60	7.9	388	6.5	<2	17.0	
	34.4	6.3	54	7.9	388	6.2	<2	14.1	
	36.1	5.8	49	7.8	389	6.1	<2	12.6	
	37.7	5.5	46	7.8	389	6.0	<2	12.0	
	39.4	4.9	41	7.8	390	5.8	<2	10.5	
	41.0	4.3	36	7.8	390	5.6	<2	8.9	
	42.6	4.0	34	7.7	391	5.5	<2	8.6	
	44.3	3.9	32	7.7	391	5.5	<2	7.8	
	45.9	3.6	30	7.7	391	5.4	<2	7.8	
	47.6	3.5	29	7.7	391	5.3	<2	7.6	
	49.2	3.4	28	7.7	392	5.3	<2	7.7	

Table 5. Physical field parameters measured at Long Lake near McCarthy, Alaska, 2008.—Continued

[**Abbreviations:** ft, foot; mg/L, milligram per liter; µS/cm at 25 C, microsiemens per centimeter at 25 degrees Celsius; C, degree Celsius; NTU, Nphelometric Turbidity Unit; –, no data]

Date	Sample depth (ft)	Dissolved oxygen (mg/L)	Oxygen saturation (percent)	pH (units)	Specific conductance (µS/cm at 25°C)	Water temperature (°C)	Turbidity (NTU)	Chlorophyll-α	Secchi disk depth (ft)
08-05-08	1.6	10.3	103	8.3	362	15.5	<2	1.1	20.9
	3.3	10.3	103	8.3	362	15.4	<2	1.1	
	4.9	10.2	102	8.3	362	15.4	<2	1.1	
	6.6	10.2	102	8.3	362	15.4	<2	.2	
	8.2	10.2	102	8.3	362	15.3	<2	.5	
	9.8	10.2	102	8.3	362	15.2	<2	.2	
	11.4	10.2	101	8.3	363	15.0	<2	.6	
	13.1	10.4	102	8.3	363	14.4	<2	1.0	
	14.8	10.5	102	8.3	363	13.9	<2	1.6	
	16.4	10.4	100	8.2	360	13.6	<2	1.7	
	18.0	10.3	98	8.2	362	13.1	<2	2.0	
	19.7	10.1	94	8.2	365	12.0	<2	2.6	
	21.3	9.9	89	8.1	368	10.9	<2	1.2	
	23.0	9.8	88	8.1	376	10.4	<2	1.5	
	24.6	9.8	86	8.0	382	9.8	<2	1.5	
	26.2	9.0	77	8.0	384	9.1	<2	1.5	
	27.9	7.8	67	7.9	385	8.7	<2	1.3	
	29.5	7.7	65	7.9	387	8.1	<2	1.6	
	31.2	7.4	62	7.9	388	7.8	<2	1.3	
	32.8	6.7	57	7.8	388	7.6	<2	2.4	
	34.4	6.3	52	7.8	388	7.5	<2	2.4	
	36.1	5.9	48	7.8	389	7.2	<2	2.2	
	37.7	4.9	39	7.8	389	6.7	<2	2.8	
	39.4	4.2	34	7.7	389	6.6	<2	3.9	
	42.6	2.8	22	7.7	391	5.9	<2	4.0	
	45.9	1.4	11	7.7	391	5.7	<2	5.1	
	49.2	1.0	8	7.6	391	5.6	<2	5.3	
09-23-08	1.6	9.8	92	8.2	360	10.4	<2	2.2	23
	3.3	9.8	92	8.2	360	10.4	<2	2.6	
	4.9	9.8	92	8.2	361	10.4	<2	3.1	
	6.6	9.8	92	8.2	360	10.4	<2	2.4	
	8.2	9.8	92	8.2	361	10.3	<2	2.1	
	9.8	9.8	92	8.2	360	10.3	<2	3.1	
	11.4	9.8	92	8.2	360	10.2	<2	2.8	
	13.1	9.8	91	8.2	361	10.2	<2	2.4	
	14.8	9.8	91	8.2	360	10.2	<2	3.0	
	16.4	9.8	91	8.2	361	10.2	<2	2.9	
	18.0	9.7	91	8.1	361	10.2	<2	2.5	
	19.7	9.6	89	8.1	361	10.2	<2	3.1	
	21.3	9.5	88	8.1	362	10.1	<2	2.4	
	23.0	9.2	86	8.1	362	10.1	<2	3.1	
	24.6	8.9	83	8.1	363	10.0	<2	2.6	
	26.2	8.8	81	8.0	364	10.0	<2	2.0	
	27.9	7.5	69	8.0	366	9.5	<2	3.0	
	29.5	4.5	40	7.8	379	8.9	<2	2.1	
	31.2	3.7	33	7.8	385	8.5	<2	2.3	
	32.8	2.9	26	7.7	388	7.9	<2	1.9	
	34.4	2.1	18	7.7	390	7.4	<2	2.0	
	36.1	1.6	14	7.7	391	7.2	<2	1.9	
	37.7	1.3	11	7.6	391	7.1	<2	1.7	

Table 5.   Physical field parameters measured at Long Lake near McCarthy, Alaska, 2008.—Continued

[**Abbreviations:** ft, foot; mg/L, milligram per liter; µS/cm at 25 C, microsiemens per centimeter at 25 degrees Celsius; C, degree Celsius; NTU, Nephelometric Turbidity Unit; –, no data]

Date	Sample depth (ft)	Dissolved oxygen (mg/L)	Oxygen saturation (percent)	pH (units)	Specific conductance (µS/cm at 25°C)	Water temperature (°C)	Turbidity (NTU)	Chlorophyll-α	Secchi disk depth (ft)
09-23-08—	39.4	0.8	7	7.6	392	6.8	<2	1.8	
Continued	41.0	0.5	4	7.6	392	6.6	<2	2.0	
	42.6	0.4	3	7.6	393	6.5	<2	2.4	
	44.3	0.3	2	7.5	393	6.5	<2	2.2	
	45.9	0.1	0	7.5	395	6.2	4.5	4.6	
	47.6	0.1	0	7.5	395	6.1	4.3	5.2	
	49.2	0.1	0	7.5	396	6.0	3.9	5.7	
	50.8	0.1	0	7.5	396	5.9	3.6	5.1	

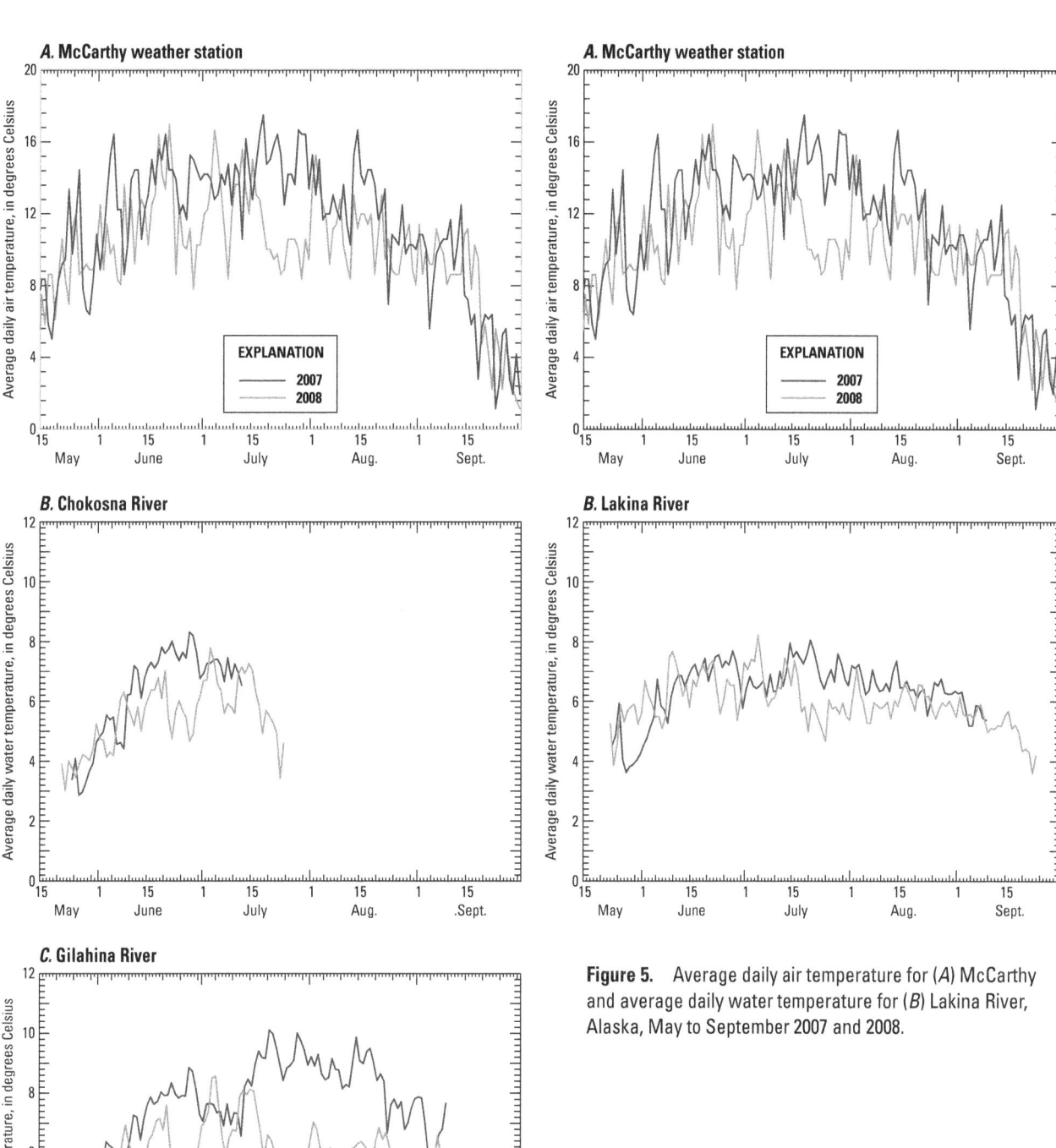

Figure 4. Average daily air temperature for (*A*) McCarthy and average daily water temperature for (*B*) Chokosna and (*C*) Gilahina Rivers, Alaska, May to September 2007 and 2008.

Figure 5. Average daily air temperature for (*A*) McCarthy and average daily water temperature for (*B*) Lakina River, Alaska, May to September 2007 and 2008.

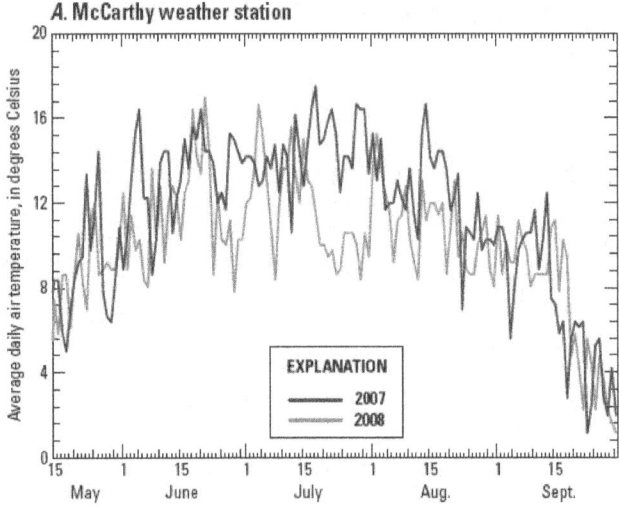

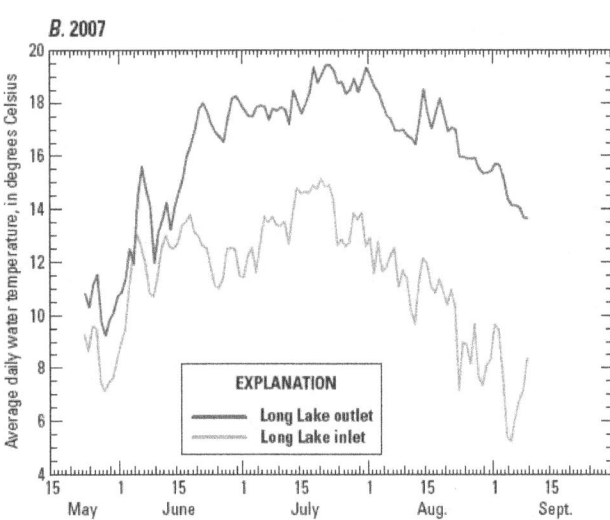

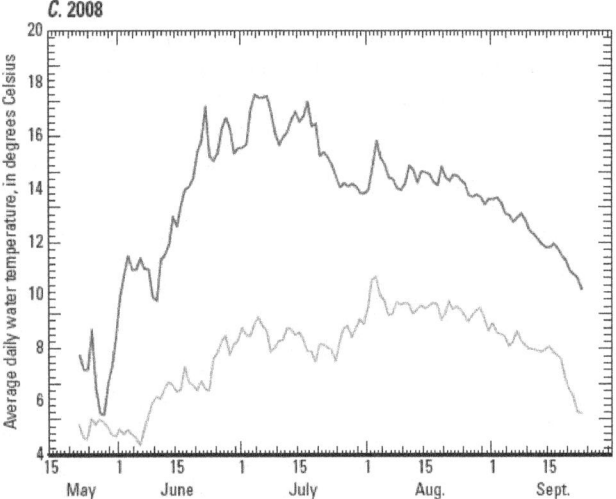

Figure 6.   Average daily air temperature for (*A*) McCarthy and average daily water temperature for (*B–C*) Long Lake outlet and inlet streams, Alaska, May to September 2007 and 2008.

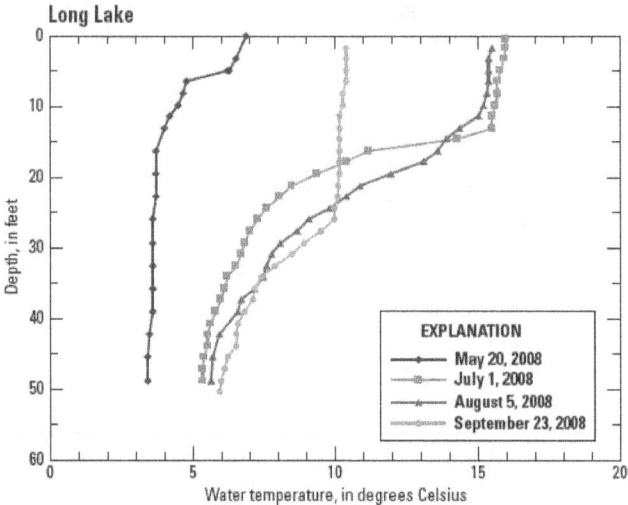

Figure 7.   Water temperature profiles for Long Lake, Alaska, May to September 2008.

Water Temperature of Long Lake Outlet—1998–2008

Mohseni and others (1998) developed a non-linear-regression model that uses a logistic function to determine average weekly stream temperatures. Inputs to the model are weekly air temperature and weekly water temperature. The model then calculates a regression equation for a particular site. Provided the resulting regression equation is accurate, one can then input long-term air temperature into the regression equation to determine long-term water temperatures. The model was used to assess water temperatures in streams and rivers in the Cook Inlet Basin of Alaska with good results (Kyle and Brabets, 2001).

The regression model for stream temperatures developed by Mohseni and others (1998) was applied to the Long Lake outlet stream to examine water temperatures for the last 10 years (1998–2008) because this stream (and Long Lake) supports a large run of sockeye salmon. Air temperature data were obtained from the climate station operated by the National Weather Service at McCarthy from 1998 to 2008. Using the air temperature data for 2007 and 2008 and the water temperature for 2007 and 2008 from the outlet stream of Long Lake, the stream temperature model was calibrated to determine the following coefficients:

a the maximum weekly stream temperature,

m the minimum weekly stream temperature,

b the air temperature at the inflection point, and

g a measure of the steepest slope of the S function.

For Mohseni's model, the Nash-Sutcliffe coefficient (NSC) (Nash and Sutcliffe, 1970) is the main determinant of goodness of fit. Root mean square error (RMSE) expresses the standard error of prediction and also is evaluated. The values for NSC and RMSE for the outlet stream of Long Lake were 0.60 and 1.83, respectively, which would indicate a fair fit of the model.

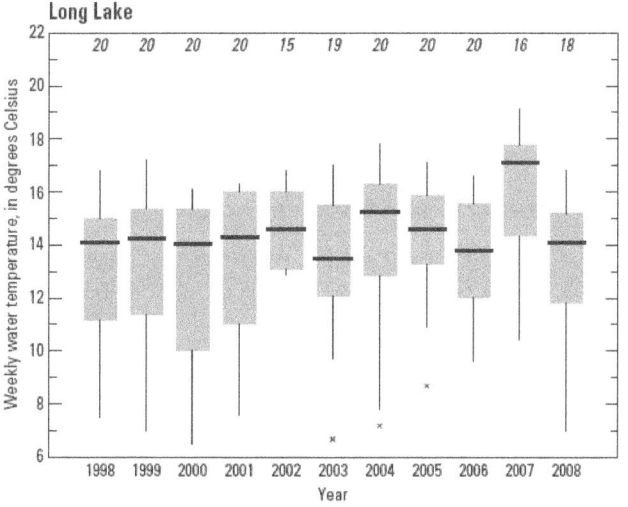

Figure 8. Distribution of simulated average weekly water temperatures from the outlet stream of Long Lake, Alaska, May through September, 1998–2008.

After the coefficients *a*, *m*, *b*, and *g* were determined, the regression model was used with the 1998–2006 air temperature data to estimate the average weekly stream temperatures for these periods. Results from the regression model (fig. 8) indicate that between 1998 and 2008, there were no significant trends in water temperature. However, 2007 experienced the highest median and highest weekly water temperatures during this 10-year period.

Specific Conductance

Specific conductance is a measure of the ability of water to conduct an electric current, and can be used to indicate the concentrations of dissolved solids, or ions, in the water. As the concentration of ions in solution increases or decreases, so does the conductance of the solution. Frequently, a statistical relation can be developed between specific conductance and the ionic components making up the dissolved solids in water. During low flow, the conductance of stream water generally is the highest, indicating a greater component of groundwater in the total flow. Groundwater has greater potential to dissolve minerals than does rainwater or snowmelt, having spent more time in contact with rocks and soil materials. Periods of relatively low specific conductance in stream water reflect

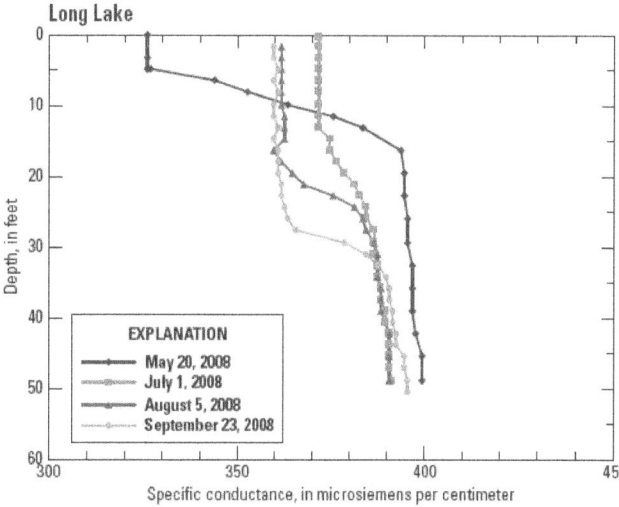

Figure 9. Specific conductance profiles for Long Lake, Alaska, May to September 2008.

runoff of rain or snowmelt, which typically contain small amounts of dissolved ions.

During the study period, measured values of specific conductance did not correlate with discharge (table 4). Most likely, the variation in concentration with discharge is attributed to the mixing of the different components of discharge, which are quantities of water from multiple sources. Specific conductance values for the Long Lake outlet would not be expected to correlate with discharge due to the mixing characteristics of Long Lake. The highest average specific conductance was that for the Chokosna River (428 µS/cm at 25 °C —seven measurements), and the lowest average specific conductance (230 µS/cm at 25 °C—six measurements) was at the Gilahina River.

At Long Lake, profiles of specific conductance were fairly uniform throughout the water column from July through September (fig. 9). Values during this period ranged from 360 to 400 µS/cm at 25 °C. The highest variability was noted in May, when values were as low as 325 µS/cm in the upper 6 ft, and then increased to near 400 µS/cm at 16 ft and remained near this level to the bottom of the lake. The lower values of conductance near the surface reflect snowmelt or rainfall runoff into the lake.

pH

The pH of water is a measure of its hydrogen-ion activity and typically ranges from 0 (acidic) to 14 (alkaline) standard units. The pH of natural river water typically ranges between 6.5 and 8.0 standard units (Hem, 1985). During the study period, measured values of pH for all sites ranged from 7.8 to 8.6 (table 4). The pH values in profiles at Long Lake varied only slightly, ranging from 7.7 to 8.4 (table 5). All values were within an acceptable range for cold-water fish growth and survival.

Dissolved Oxygen

The dissolved-oxygen concentration in a stream is controlled by several factors, including water temperature, air temperature and atmospheric pressure, hydraulic characteristics of the stream, photosynthetic or respiratory activity of stream biota, and the quantity of organic matter present (Hem, 1985). Salmon and other species of fish require well-oxygenated water at every stage in their life history, as do many forms of aquatic invertebrates. Young fish tend to be more susceptible to oxygen deficiencies than adults. Measurements of dissolved oxygen at all stream sites during the study period ranged from 6.1 to 13.2 mg/L (table 4). The range of values was nearly identical at all sites.

At Long Lake, the levels of dissolved-oxygen concentration varied considerably with depth throughout the measurement period in 2008 (fig. 10). In May, dissolved-oxygen concentration was approximately 9 to 10 mg/L from the surface to 6 ft. A strong stratification exists between 6 and 13 ft where dissolved-oxygen concentration decreases to less than 4 mg/L. In July, concentrations were near 10 mg/L from the surface to 16 ft and then increased to 12 mg/L from 16 to 25 ft. This increased dissolved-oxygen concentration between 16 and 33 ft indicates the presence of phytoplankton that had settled to the top of the density barrier of the thermocline and were actively photosynthesizing. Dissolved-oxygen concentrations then steadily decreased to about 3 mg/L at the bottom. In August, concentrations were near 10 mg/L at the surface to 26 ft and then decrease to about 1 mg/L at the bottom, showing a profile similar to the July profile. In September, concentrations were between 9 and 10 mg/L from 0 to 23 ft. A strong stratification exists from 26 to 36 ft as concentrations decrease to 2 mg/L and then continued to decline to 0 mg/L at the lake bottom. The late summer stratification prevents the replenishment of oxygen below the thermocline where oxygen is consumed by respiration and decomposition.

Alkalinity

Alkalinity is a measure of the capacity of the substances dissolved in water to neutralize acid. In most natural waters, alkalinity is produced mainly by bicarbonate and carbonate ions, which are formed when carbon dioxide or carbonate rocks dissolve in water (Hem, 1985). The lowest alkalinity values were at the Gilahina River and ranged from 63 to 96 mg/L (table 6); the highest alkalinity values were measured at the Long Lake inlet and outlet streams, and ranged from 92 to 222 mg/L and 117 to 164 mg/L, respectively. The range of pH measured at all sites indicates that all of the alkalinity can be attributed to dissolved bicarbonate (Hem, 1985). Alkalinity concentrations at all sites would be considered average to high and would indicate a high buffering capacity.

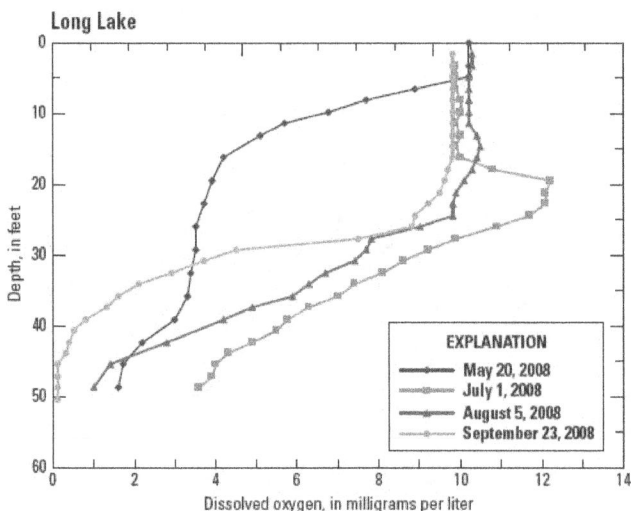

Figure 10.   Dissolved oxygen profiles for Long Lake, Alaska, May to September 2008.

Major Dissolved Constituents and Iron and Manganese

Major dissolved constituents in streams consist of inorganic minerals derived primarily from soil and rock weathering. Dissolved cations that constitute a majority of the dissolved solids content in natural waters are calcium, magnesium, sodium, and potassium; the major anions are usually represented by sulfate, chloride, fluoride, nitrate and those making up the alkalinity (carbonate and bicarbonate) (Hem, 1985). Streams draining basins with rocks and soils containing insoluble minerals tend to have lower concentrations of dissolved solids than streams draining basins with easily dissolved minerals. Analyses of the water samples collected in this study indicated that total dissolved solid concentrations are highest (average of 291 mg/L) in the Chokosna River and lowest (average of 141 mg/L) in the Gilahina River (table 6). Dissolved solids concentrations in samples from Long Lake inlet and outlet averaged 187 and 211 mg/L, respectively. Concentrations of dissolved solids such as these are representative of basins containing soils and rocks that are easily dissolved.

Calcium and magnesium are common alkaline-earth metals that are essential elements in plant and animal nutrition. Both elements are major anion components in most natural waters (Hem, 1985). In the samples collected in this study, concentrations of calcium ranged from 28 to 86 mg/L, and those of magnesium ranged from 2.8 to 12.0 mg/L (table 6). Concentrations of these constituents tended to be higher at Chokosna River and Long Lake sites than at the Gilahina River. Sodium and potassium are both present in most natural waters, but usually in low concentrations in rivers. Sodium concentrations ranged from 1.8 to 9.2 mg/L and potassium concentrations ranged from 0.25 to 1.02 mg/L at all sites (table 6).

Table 6. Concentrations of major ions and trace elements iron and manganese in water samples collected from streams along McCarthy Road, Alaska, 2007 and 2008.

[All values in milligrams per liter unless otherwise noted. Values in parentheses with date for Long Lake are measurement depth. **Abbreviations**: µg/L, micrograms per liter; ft, foot; E, estimated; <, less than]

Date	Time	Depth (ft)	Alkalinity	Bicarbonate	Calcium	Chloride	Total dissolved solids	Fluoride	Iron (µg/L)	Magnesium	Manganese (µg/L)	Potassium	Silica	Sodium	Sulfate
							Lakina River								
05-22-07	1530	–	92	112	42.6	0.36	147	E0.08	23	6.6	3.6	0.80	5.4	3.0	41.6
10-02-07	1715	–	106	129	53.4	.35	211	E.11	<8	8.5	1.1	.57	6.2	3.4	63.2
							Long Lake Inflow								
05-22-07	1000	–	92	112	28.9	0.73	119	E0.09	64	4.8	12.6	0.54	7.0	3.7	2.8
07-30-07	0945	–	222	271	61.0	1.71	260	.10	464	10.3	476	.25	10.6	9.2	1.9
05-21-08	1030	–	140	171	44.2	.94	174	E.09	33	7.1	7.3	.60	7.6	6.0	6.2
07-01-08	1000	–	169	206	49.9	1.21	198	E.10	37	8.0	10.3	.46	8.0	7.4	4.8
08-06-08	0915	–	133	162	41.6	.82	172	E.09	39	6.8	7.7	.50	8.3	5.3	4.1
09-24-08	1020	–	171	209	54.6	1.12	199	E.10	32	8.5	11.8	.62	8.9	7.4	6.8
							Long Lake Outflow								
05-22-07	1210	–	117	143	41.3	0.69	157	E0.07	20	7.1	4.8	0.69	6.1	4.1	16.0
07-30-07	1900	–	154	188	49.2	.86	204	E.08	E3	9.5	1.4	.82	5.6	5.3	26.0
05-20-08	1730	–	154	188	54.0	.88	220	E.10	10	9.2	3.3	.86	6.8	5.1	27.9
07-01-08	1125	–	164	202	56.9	.93	232	E.10	<8	10.2	2.2	.89	7.0	5.5	31.5
08-05-08	1644	–	164	200	54.7	.92	225	E.09	<8	9.8	.8	.89	6.2	5.2	30.0
09-23-08	1745	–	160	195	57.3	.92	226	E.10	<8	9.8	.9	.86	6.7	5.7	28.4
							Gilahina River								
05-23-07	0900	–	63	76	28.2	0.20	92	<0.10	52	2.8	1.2	0.73	6.0	1.8	16.8
08-01-07	1200	–	91	111	41.8	.13	166	<.10	8	4.5	.5	.39	7.9	2.9	37.9
10-03-07	1600	–	88	107	41.2	.23	148	<.12	9	4.4	.9	.64	8.2	2.9	31.6
05-19-08	1810	–	82	100	31.8	.16	124	<.12	38	3.1	1.6	.62	6.8	1.9	18.0
06-30-08	1730	–	79	97	35.9	E.11	145	E.07	9	3.7	.8	.33	7.1	2.6	30.8
09-24-08	1600	–	96	117	46.8	.17	174	<.12	12	4.7	1.4	.42	8.2	2.9	38.1
							Chokosna River								
05-23-07	1100	–	64	78	35.8	0.20	190	<0.10	29	4.8	4.2	0.77	4.0	3.1	44.7
07-29-07	1400	–	121	148	78.8	.22	334	E.07	64	10.5	.9	.47	5.9	6.2	134
10-03-07	1240	–	112	137	80.6	.25	326	.13	<8	11.1	.8	.43	6.4	6.5	138
05-19-06	1545	–	92	113	52.2	.26	220	E.08	16	7.0	2.0	.68	5.8	4.6	68.0
06-30-08	1540	–	110	134	69.5	.17	294	E.09	<8	9.4	.9	.45	5.5	5.8	116
08-04-08	1635	–	110	134	70.4	.17	310	E.09	<8	10.5	1.2	.46	5.7	5.5	122
09-24-08	1700	–	123	150	86.5	.24	366	E.07	<8	12.0	.5	.50	6.2	6.6	145
							Long Lake								
05-20-08	1300	3	148	181	52.2	0.86	207	E0.09	11	9.0	1.8	0.85	6.8	4.7	25.9
05-20-08	1310	33	180	220	65.1	1.00	251	.12	<8	11.4	.5	.98	7.7	6.0	32.1
07-01-08	1415	10	169	206	58.7	1.00	236	E.10	<8	10.3	.8	.91	7.0	5.5	31.4
07-01-08	1425	39	174	212	59.4	1.00	240	E.11	<8	10.4	6.1	.94	8.3	5.4	32.6
08-05-08	1210	15	169	199	55.5	.98	219	E.10	<8	9.8	E.3	.93	6.4	5.4	29.5
08-05-08	1230	39	178	215	60.3	1.05	234	E.10	<8	10.6	2.1	1.02	8.0	5.6	31.5
09-23-08	1840	20	167	197	57.6	1.39	225	E.09	E6	9.7	.7	.89	6.7	5.6	28.3
09-23-08	1850	39	178	213	62.9	1.05	248	E.09	<8	10.5	2.7	1.00	9.2	5.8	31.3

Bicarbonate (HCO_3) was the dominant anion at all sites except the Chokosna River, where sulfate was also a dominant anion. Concentrations of bicarbonate were the highest at the Long Lake sites and more variable at the Gilahina and Chokosna Rivers. Silica and sulfate, which are dissolved from rocks and soils, are the next most abundant anions, with concentrations ranging from 4.0 to 10.6 mg/L for silica, and from 1.9 to 145 mg/L for sulfate. Chloride and fluoride concentrations were less than or equal to 1.71 and 0.13 mg/L, respectively at all sites (table 6).

Iron is dissolved from many rocks and soils and is an essential element in the metabolism of animals and plants. Iron in drinking water does not pose a health threat provided concentrations are less than 300 µg/L. Concentrations at all sites were less than 100 µg/L (table 6) with the exception of one sample at the Long Lake inlet, in which the concentration of iron was 464 µg/L. The chemistry of manganese (Mn) is similar to that of iron and Mn concentrations should generally be less than 50 µg/L. Concentrations of manganese were less than 13 µg/L in all but one sample, from the Long Lake inlet, with a concentration of 476 µg/L.

A trilinear diagram (Hem, 1985) was used to plot the concentrations of dissolved major ions in the water samples. This diagram permits the chemical composition of multiple samples to be represented on a single graph, and facilitates classification of the sample chemistry. On the basis of analyses of the samples collected during this study, the water of the Chokosna River can be classified as a calcium bicarbonate sulfate water (fig. 11). The water of the remaining sites would be classified as a calcium bicarbonate water.

Nutrients and Dissolved Organic Carbon

Nitrogen (N) is present in the crustal rocks of the earth and in the atmosphere. In its reduced or organic forms, it is converted by soil bacteria into nitrite (NO_2) and nitrate (NO_3). Biological fixation is accomplished by blue-green algae and certain related organisms that have the capacity for photosynthesis. Nitrogen is an important water-quality constituent largely due to its role as a component of the chlorophyll in plants and thus is essential for primary productivity in lakes, streams, and rivers. In aquatic ecosystems, N commonly occurs in three ionic forms: ammonium (NH_4), (NO_2), and (NO_3). In the laboratory, NH4 is analyzed as ammonia (NH_3); thus N concentrations are reported as: total and dissolved NH_3 plus organic N; dissolved NH_3; dissolved NO_2 plus NO_3; and dissolved NO_2. Nitrate is generally more abundant than NO_2 in natural waters because NO_2 readily oxidizes to NO_3 in oxygenated water (Hem, 1985). Total NH_3 plus organic N concentrations for whole water samples represent the sum of biologically derived organic N compounds, plus any NH_3 present. Nitrate and NO_2 are oxidized forms of inorganic N that make up most of the dissolved N in well-oxygenated streams such as those along the McCarthy Road. The dissolved concentrations represent the NH_4 or NO_2 plus NO_3 in solution and associated with material capable of passing through a 0.45-µm-pore filter.

All concentrations of the various N forms were less than 1 mg/L (table 7). Due to its toxicity to freshwater aquatic organisms, the U.S. Environmental Protection Agency (USEPA; 1976) suggests a limitation of 0.02 mg/L of NH_3 as un-ionized NH_3 for waters to be suitable for fish propagation. Concentrations of NH_3 (both ionized and unionized) were all below this level at all sites.

Phosphorus (P) is a rather common element in igneous rock. It is fairly abundant in sediments, but at concentrations in natural waters normally no more than a few tenths of a milligram per liter. Phosphorus is an element vital to all forms of aquatic biota because it is involved in the capture and transfer of chemical energy, and it is an essential element in nucleic acids (Gaudy and Gaudy, 1988). It occurs as organically bound P or as phosphate (PO_4^{3-}). Elevated concentrations of P in water are not considered toxic to human or aquatic life. Elevated concentrations can, however, stimulate the growth of algae in lakes and streams. Phosphorus concentrations are reported as total P and dissolved orthophosphate (PO_4). Total PO_4^{3-} concentrations represent the P in solution, associated with colloidal material, and contained in or attached to biotic and inorganic particulate matter. Dissolved concentrations are determined from the filtrate that passes through a filter with a nominal pore size of 0.45 µm. The PO_4 ion is a significant form of P because it is directly available for metabolic use by aquatic biota. Concentrations of total P, dissolved P and PO_4 were typically low, with values near or below minimum detection levels in nearly all samples (table 7).

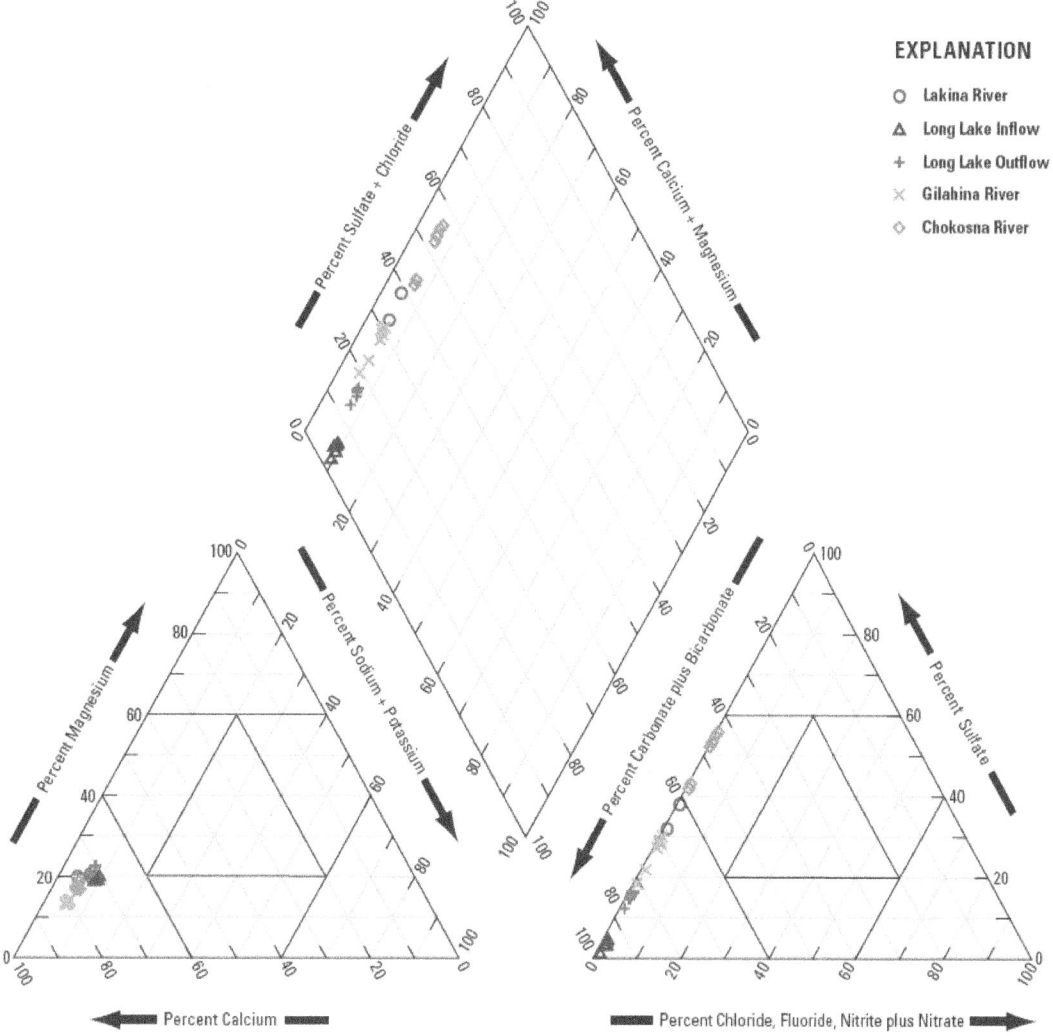

Figure 11. Chemical composition of water samples collected from sites along McCarthy Road, Alaska, 2007 and 2008.

Phytoplankton (algae) can assimilate only three of the several different nutrient (N and P) compounds commonly detected in aquatic ecosystems. The only forms of N that phytoplankton can use for growth are NO_2, NO_3, and NH_4 (inorganic nitrogen), whereas PO_4 is the only form of P that phytoplankton can use for growth (Horne and Goldman, 1994). An N:P ratio greater than 10 (by weight) generally indicates that P is the limiting nutrient, whereas a ratio less than 10 indicates that N is limiting (Horne and Goldman, 1994). For Long Lake, N:P ratios ranged from 6 to 12 above the thermocline and in the euphotic zone (table 7).These N:P ratios suggest that at the beginning of open water (May) and at the end of summer (September) phosphorus is the limiting nutrient but during the summer months (July and August) nitrogen is the limiting nutrient.

Dissolved organic carbon (DOC) is commonly a major component of organic matter in aquatic ecosystems. DOC is defined as organic carbon in the filtrate (dissolved and colloidal phases) that passes through a 0.45-μm pore-size filter (Aiken and Cotsaris, 1995). Generally, DOC is in greater abundance than particulate organic carbon, accounting for about 90 percent of the total organic carbon of most waters (Aiken and Cotsaris, 1995). At the McCarthy Road sites, the highest concentrations of DOC—6.3–11.7 mg/L—were detected at the inlet stream to Long Lake. This watershed is comprised primarily of wetlands; thus, DOC concentrations would be expected to be higher than the other streams. At the Gilahina and Chokosna Rivers, DOC concentrations did exhibit some seasonality and were highest in May 2007 and 2008, reflecting spring runoff.

Table 7. Concentrations of nutrients and dissolved organic carbon in water samples collected along McCarthy Road, Alaska, 2007 and 2008.

[All values in milligrams per liter unless otherwise noted. **Abbreviations:** ft, foot; N:P, nitrogen:phosphorus; <, less than; E, estimated; –, not applicable]

Date	Time	Depth (ft)	Dissolved ammonia nitrogen (NH$_4$)	Dissolved nitrogen (NH$_4$+Org)	Total nitrogen (NH$_4$+Org)	Dissolved nitrogen (NO$_2$+NO$_3$)	Dissolved nitrogen (NO$_2$)	Dissolved phosphorus	Dissolved ortho-phosphorus	Total phosphorus	Dissolved organic carbon	N:P ratio
Lakina River												
05-22-07	1530	–	<0.020	0.26	0.21	0.155	E0.002	<0.006	<0.006	0.106	3.3	–
10-02-07	1715	–	<.020	<.14	<.14	.197	<.004	<.006	.004	.021	1.2	–
Long Lake Inflow												
05-22-07	1000	–	<0.020	0.43	0.30	<0.016	E0.001	E0.004	<0.006	E0.006	11.7	–
07-30-07	0945	–	<.020	.31	.29	<.016	<.002	.006	E.005	.009	10.3	–
05-21-08	1030	–	<.020	.26	.19	.037	E.001	<.006	E.004	E.005	6.3	–
07-01-08	1000	–	<.020	.26	.22	.02	<.002	E.003	<.006	<.008	8.8	–
08-06-08	0915	–	<.020	.32	.31	<.016	<.002	<.006	.006	E.007	11.5	–
09-24-08	1020	–	<.020	.21	.19	E.009	<.002	<.006	E.004	E.005	6.9	–
Long Lake Outflow												
05-22-07	1210	–	<0.020	0.28	0.23	<0.016	E0.001	<0.006	<0.006	0.011	5.8	–
07-30-07	1900	–	<.020	.19	.30	<.016	<.002	E.005	E.004	E.007	5.0	–
05-20-08	1730	–	<.020	.17	.21	.024	<.002	E.004	E.003	.013	3.6	–
07-01-08	1125	–	<.020	.28	.20	<.016	<.002	E.005	<.006	.008	4.1	–
08-05-08	1644	–	<.020	.19	.21	<.016	<.002	E.005	E.005	E.004	4.8	–
09-23-08	1745	–	<.020	.20	.19	<.016	<.002	E.004	E.003	E.007	4.3	–
Gilahina River												
05-23-07	0900	–	<0.020	0.21	0.19	0.038	E0.001	<0.006	<0.006	0.027	3.9	–
08-01-07	1200	–	<.020	E.07	<.10	.103	<.002	E.003	E.004	<.008	1.7	–
10-03-07	1600	–	<.020	E.11	E.09	.144	<.002	<.006	E.003	.011	2.2	–
05-19-08	1810	–	<.020	E.11	E.14	.046	<.002	<.006	.009	.019	2.9	–
06-30-08	1730	–	<.020	E.10	<.14	.122	<.002	<.006	<.006	.021	2.1	–
09-24-08	1600	–	<.020	E.11	E.13	.116	<.002	<.006	E.004	E.007	2.0	–
Chokosna River												
05-23-07	1100	–	<0.020	0.26	0.17	0.037	E0.001	E0.003	<0.006	0.079	3.5	–
07-29-07	1400	–	<.020	E.06	<.10	.098	<.002	E.003	E.004	E.007	.9	–
10-03-07	1240	–	<.020	E.07	<.10	.145	<.002	<.006	E.003	.013	1.1	–
05-19-06	1545	–	<.020	E.12	E.14	.055	E.001	<.006	<.006	.028	2.9	–
06-30-08	1540	–	<.020	E.09	<.14	.095	<.002	<.006	<.006	.041	1.3	–
08-04-08	1635	–	<.020	E.08	E.09	.132	<.002	<.006	E.006	.079	1.5	–
09-24-08	1700	–	<.020	E.08	E.12	.144	<.002	<.006	E.003	E.005	1.1	–

Table 7. Concentrations of nutrients and dissolved organic carbon in water samples collected along McCarthy Road, Alaska, 2007 and 2008.—Continued

[All values in milligrams per liter unless otherwise noted. **Abbreviations:** ft, foot; N:P, nitrogen:phosphorus; <, less than; E, estimated; <, less than; –, not applicable]

Date	Time	Depth (ft)	Dissolved ammonia nitrogen (NH_4)	Dissolved nitrogen (NH_4+Org)	Total nitrogen (NH_4+Org)	Dissolved nitrogen (NO_2+NO_3)	Dissolved nitrogen (NO_2)	Dissolved phosphorus	Dissolved ortho phosphorus	Total phosphorus	Dissolved organic carbon	N:P ratio
						Long Lake						
05-20-08	1300	3	<0.020	0.16	0.29	E0.011	E0.001	<0.006	E0.003	0.019	3.7	10
05-20-08	1310	33	<.020	E.13	.18	.085	<.002	E.003	E.004	.012	3.3	26
07-01-08	1415	10	<.020	.20	.17	<.016	<.002	E.004	<.006	E.005	4.0	6
07-01-08	1425	39	<.020	.17	.23	.04	<.002	E.004	<.006	.012	3.8	10
08-05-08	1210	15	<.020	.18	.23	<.016	<.002	<.006	E.006	.009	5.3	6
08-05-08	1230	39	<.020	.19	.56	.033	<.002	E.004	.007	E.007	3.9	8
09-23-08	1840	20	<.020	.18	.19	<.016	<.002	<.006	E.003	.011	4.2	12
09-23-08	1850	39	<.020	.18	.19	.121	<.002	.006	E.006	.015	3.4	24

Trace Elements in Streambed Sediments

Streambed sediment samples collected at 5 stream sites along the McCarthy Road were analyzed for 33 trace elements and organic carbon content (tables 3 and 8). Aquatic-life criteria have not been established for most of the trace elements; therefore, to provide a general comparison, the concentrations of trace elements in samples from the McCarthy Road sites were compared to concentrations in the USGS NAWQA program data set, which consists of about 1,000 samples collected throughout the contiguous United States, Alaska, and Hawaii. Of these samples, about 250 represent reference or forested areas (Horowitz and Stephens, 2008). The concentrations of trace elements at the McCarthy Road stream sites were also compared to concentrations in streambed sediments collected from the south side of Denali National Park and Preserve (representative of another interior Alaska location with similar climate) in 1999 and 2000 (Brabets and Whitman, 2002).

The range in concentrations of most of the trace elements in sediment samples collected at the sites along the McCarthy Road were similar to the medians of the NAWQA reference/forested sites, with the exception of higher concentrations of arsenic, chromium, copper, strontium, and vanadium at the McCarthy Road sites (table 9). Concentrations of most trace elements detected in streambed samples from the south side of Denali National Park and Preserve were similar to concentrations in the McCarthy Road samples. Notable exceptions were higher concentrations of copper, manganese, and strontium in the McCarthy Road samples.

The focus in the literature on criteria for streambed sediments has been limited to nine trace elements: arsenic, cadmium, chromium, copper, lead, mercury, nickel, selenium, and zinc. Guidelines have been established for these elements in unsieved streambed sediment. Because trace-element samples for this study are from sediments finer than 0.063 mm, in which concentrations tend to be greatest, comparisons with the guidelines may overestimate the effects on aquatic organisms (Deacon and Stephens, 1998). However, it was still felt that it would be acceptable to compare the concentrations from the finer than 0.063-mm size fraction to the published guidelines for this study.

The Canadian Council of Ministers of the Environment (1999) has established guidelines for the protection of aquatic life for some trace elements in unsieved streambed sediment. These guidelines use an assessment value called the probable effect level (PEL), which is the concentration above which adverse effects on aquatic organisms are expected to occur frequently (table 10). MacDonald and others (2000) proposed sediment quality guidelines for eight trace elements, and Van Derveer and Canton (1997) proposed guidelines for selenium (table 10). MacDonald and others (2000) use a value called the probable effect concentration (PEC) and assume a 1 percent organic carbon concentration. The PEC is the concentration above which toxicity is likely. Van Derveer and Canton (1997) suggested a value of 4 µg/g for selenium that is comparable to the PEC. Concentrations greater than the PEL for one or more trace elements were noted in samples from several sites: for arsenic (Lakina River, Long Lake inflow), chromium (Gilahina River), and nickel (Lakina River, Long Lake inflow, Gilahina River, and Chokosna River). Concentrations greater than the PEC for these elements were detected only at the Lakina River and Gilahina River. These concentrations reflect the local geology of this natural area rather than anthropogenic sources.

MacDonald and others (2000) developed a Mean PEC Quotient, to describe the toxicity of the combined trace element concentrations. The Mean PEC Quotient is computed by summing the concentrations of all the trace elements analyzed and dividing by the number of elements. MacDonald and others (2000) determined that sediments with mean PEC quotients less than 0.5 accurately predicted the absence of toxicity in 83 percent of the samples they examined. Mean PEC quotients greater than 0.5 accurately predicted toxicity in 85 percent of the samples. Comparison of the concentrations of the trace elements with the percent organic carbon and mean PEC Quotient provides some insights about the bioavailability of these trace elements. The concentration of organic carbon in sediment is used to indicate the concentration of organic matter. The ability of organic matter to concentrate some trace elements in stream sediment is well recognized (Gibbs, 1973; Horowitz, 1991), and this ability varies with the type of organic matter. For example, complexation by organic matter, such as humic and fulvic acids, has generally been thought to reduce bioavailability of certain metals (Spacie and Hamelink, 1985, Newman and Jagoe, 1994). Results of studies by Winner (1985) and Decho and Luoma (1994), however, suggest that organic carbon compounds may in some cases enhance uptake of certain trace elements. The organic carbon content in three of the five streambed sediment samples (Long Lake Inlet, Long Lake Outlet, and Gilahina River) was greater than 1 percent (3.5, 5.4, and 1.7, respectively), and the corresponding mean PEC quotient was less than 0.50 at these three sites (table 8). As MacDonald and others (2000) noted, sites containing relatively low concentrations of organic carbon have higher potential toxicity. When normalized to percent organic carbon, concentrations of arsenic at the Lakina River and concentrations of chromium and nickel at the Chokosna River were above the PEC level.

Table 8. Concentrations of trace elements and percentage of carbon measured in streambed sediments collected from sites along McCarthy Road, Alaska, 2007.

[Site No.: Locations shown in figure 2. Values of carbon, aluminum, sodium, iron, titanium, calcium, magnesium, phosphorus, and potassium in percent; all other values in micrograms per gram. Abbreviations: PEC, probable effect concentration; M, measured; I-Carbon, inorganic carbon; O-Carbon, organic carbon; <, less than]

Site No.	Site name	Date	Time	Carbon(total)	Aluminum	Antimony	Arsenic	Barium	Beryllium	Bismuth	Cadmium
1	Lakina River	10-02-07	1445	4.3	5	2.4	46	550	0.9	M	0.4
2	Long Lake Inflow	07-30-07	0930	4.8	5.6	.8	29	740	1.2	M	.2
3	Long Lake Outflow	07-30-07	1800	5.4	5.4	1.1	9	490	1.0	M	.6
4	Gilahina River	08-01-07	0900	2.2	6.2	.9	13	550	1.1	M	.6
5	Chokosna River	07-29-07	1030	1.8	6.8	1.0	16	920	1.1	M	.7

Site No.	Site name	Date	Time	Cerium	Chromium	Cobalt	Copper	Gallium	Sulfur	Sodium
1	Lakina River	10-02-07	1445	27	66	13	49	11	0.07	1.1
2	Long Lake Inflow	07-30-07	0930	30	93	20	61	14	.06	1.4
3	Long Lake Outflow	07-30-07	1800	30	79	13	40	12	.24	1.4
4	Gilahina River	08-01-07	0900	45	170	28	90	15	.08	1.4
5	Chokosna River	07-29-07	1030	33	85	22	97	15	.20	1.0

Site No.	Site name	Date	Time	Iron	Lanthanum	Lead	Lithium	Manganese	Mercury	Molybdenum
1	Lakina River	10-02-07	1445	3.1	16	9	25	490	0.17	2.6
2	Long Lake Inflow	07-30-07	0930	6.7	15	6	26	10,000	.09	.8
3	Long Lake Outflow	07-30-07	1800	3.1	16	6	37	500	.15	.9
4	Gilahina River	08-01-07	0900	5.8	22	7	18	1,200	.09	1.4
5	Chokosna River	07-29-07	1030	5.8	18	16	25	1,300	.06	2.5

Site No.	Site name	Date	Time	Nickel	Niobium	Scandium	Selenium	Silver	Strontium	Thallium
1	Lakina River	10-02-07	1445	39	7	12	1.8	<1.0	690	M
2	Long Lake Inflow	07-30-07	0930	41	10	17	.3	<1.0	330	M
3	Long Lake Outflow	07-30-07	1800	32	9	14	3.6	<1.0	350	M
4	Gilahina River	08-01-07	0900	63	11	28	1.2	<1.0	370	M
5	Chokosna River	07-29-07	1030	36	9	20	1.7	<1.0	340	M

Site No.	Site name	Date	Time	Thorium	Titanium	Vanadium	Yttrium	Uranium	I-Carbon
1	Lakina River	10-02-07	1445	3	0.32	120	21	2.6	3.4
2	Long Lake Inflow	07-30-07	0930	4	.46	150	20	1.7	1.2
3	Long Lake Outflow	07-30-07	1800	3	.43	120	20	2.2	1.0
4	Gilahina River	08-01-07	0900	5	.55	230	31	2.2	.5
5	Chokosna River	07-29-07	1030	5	.46	190	26	2.7	1.1

Site No.	Site name	Date	Time	Calcium	Cesium	Magnesium	Phosphorus	Potassium	O-Carbon	Mean PEC quotient
1	Lakina River	10-02-07	1445	9.4	2.2	1.3	0.09	0.96	0.89	0.51
2	Long Lake Inflow	07-30-07	0930	4.7	2.1	1.7	.10	.94	3.5	.12
3	Long Lake Outflow	07-30-07	1800	4.7	2.0	1.1	.13	.87	5.4	.06
4	Gilahina River	08-01-07	0900	5.1	1.7	2.7	.15	.79	1.7	.32
5	Chokosna River	07-29-07	1030	4.3	2.6	1.6	.13	1.30	0.7	.59

Table 9.   Comparison of trace element concentrations measured in streambed sediments from the U.S. Geological Survey National Water-Quality Assessment program, and range of concentrations in sediment samples from streams along McCarthy Road, and 16 sites on the south side of Denali National Park and Preserve, Alaska.

[Values for aluminum, iron, and titanium in percent; all other values in micrograms per gram. **Abbreviations:** NAWQA, National Water-Quality Assessment; M, measured; –, no data; <, less than]

Trace element	Median of values for NAWQA reference forested sites (number of samples varies between 241 and 262)	Range in concentration from sites located along McCarthy Road	Range in concentration from sites located on the south side of Denali National Park	Trace element	Median of values for NAWQA reference forested sites (number of samples varies between 241 and 262)	Range in concentration from sites located along McCarthy Road	Range in concentration from sites located on the south side of Denali National Park
Aluminum	6.5	5 – 6.8	5.1 – 7.4	Manganese	955	490 – 10,000	240 – 1,600
Antimony	.7	0.8 – 2.4	0.1 – 4.2	Mercury	.07	0.06 – 0.17	<0.02 – 0.24
Arsenic	7.0	9 – 46	1.7 – 88	Molybdenum	1	0.8 – 2.6	0.5 – 1.6
Barium	470	490 – 920	480 – 1,400	Neodymium	–	–	15 – 180
Beryllium	2.0	0.9 – 1.2	1.2 – 4.0	Nickel	26	32 – 63	2 – 130
Bismuth	–	<1 (M)	<1	Niobium	–	7 – 11	7 – 23
Cadmium	.4	0.2 – 0.7	0.1 – 1.9	Scandium	12	12 – 28	2 – 17
Cerium	70	27 – 45	32 – 400	Selenium	.7	0.3 – 3.6	0.1 – 5.2
Chromium	63	66 – 170	3.0 – 220	Silver	.2	<1.0	0.1 – 0.8
Cobalt	15	13 – 28	1.0 – 26	Strontium	140	330 – 690	69 – 340
Copper	26	40 – 97	3 – 64	Tantalum	–	–	<1 – 2
Europium	1.0	–	1 – 2	Thallium	–	M	<1 – 1
Gallium	16	11 – 15	13 – 22	Thorium	12.0	3 – 5	6 – 63
Gold	–	–	1	Tin	2.5	–	2 – 5
Holmium	–	–	1 – 3	Titanium	.37	0.32 – 0.55	0.1 – 0.5
Iron	3.6	3.1 – 6.7	0.4 – 4.7	Uranium	3.9	1.7 – 2.7	2.7 – 22
Lanthanum	–	15 – 22	16 – 190	Vanadium	84	120 – 230	5 – 170
Lead	24	6 – 16	13 – 76	Yttrium	–	20 – 31	13 – 75
Lithium	–	18 – 37	37 – 75	Zinc	110	92 – 160	16 – 170

Table 10.   Summary of streambed-sediment quality guidelines for nine priority trace elements in streambed sediment at selected sites along McCarthy Road, Alaska.

[**PEL:** Values from Canadian Council of Ministers of the Environment, 1995. **PEC:** Values from MacDonald and others, 2000. Selenium values from Van Derveer and Canton, 1997. **Abbreviations:** NAWQA, National Water-Quality Assessment; PEL, probable effect level; PEC, probable effect concentration; NG, no guidelines; µg/g, microgram per gram, dry weight]

Constituent	NAWQA national median (µg/g)	PEL (µg/g)	PEC (µg/g)	Sediment sample concentration	
				Greater than PEL	Greater than PEC
Arsenic	6.35	17	33	Lakina River Long Lake Inflow	Lakina River
Cadmium	.4	3.53	4.98		
Chromium	62	90	111	Gilahina River	Gilahina River
Copper	26	197	149		
Lead	24.3	91.3	128		
Mercury	.06	.486	1.06		
Nickel	25	36	48.6	Lakina River Long Lake Inflow Gilahina River Chokosna River	Gilahina River
Selenium	.7	NG	4		
Zinc	110	315	459		

Biological Characteristics

Lake Zooplankton

Zooplankton samples were collected four times in 2008—May, July, August, and September—from the sampling location on Long Lake. The total number of zooplankton individuals varied widely, ranging from a low of 97 in July to a high of 1,037 in September (table 11). For the two groups enumerated, cladoceran and copepod, equal numbers of both groups were found in May. The cladoceran group dominated in July and August, and the copepod group dominated in September. In the cladoceran group, cyclopoid1 was the dominant family and in the copepod group, cyclopoid2 was the dominant family in May, July, and August and nauplii the dominant family in September.

Macroinvertebrates

Samples for macroinvertebrates and algae were collected at similar times in the runoff season—in late July 2007 and early August 2008. In 2007, high flow (932 ft^3/s) at the Lakina River prevented any sample collection and the site was not sampled until October 2007. Flows at the other sites were 0.2, 4, 53, and 38 ft^3/s at the Long Lake inlet stream, outlet stream, Gilahina River and Chokosna River, respectively. In 2008, flows were too high at the Gilahina and Chokosna Rivers for sample collection and samples were not collected at these sites until September 2008. The flow at the Long Lake inlet stream was relatively high at the time of sampling (14 ft^3/s) and was sampled again in September 2008 at a lower flow (3.6 ft^3/s). Characteristics of each stream reach were documented with a set of notes and photographs, depicting major geomorphic channel units such as sloughs, riffles, rapids, and other features such as beaver dams or woody debris.

The channel gradients of the Lakina, Gilahina, and Chokosna Rivers are relatively steep, as evidenced by high velocities, ranging from 3.4 to 3.8 ft/s. In contrast, the gradients of the two Long Lake streams are relatively flat as evidenced by the low average velocities at each site, 1.1 and 1.2 ft/s, respectively. The Long Lake sites represent pristine complex habitats characterized by pools and woody debris in the inlet stream and grasses and overhanging shrub in the outlet stream (figs. 12–13). Substrate in both streams is dominated by sand, small cobbles and gravel and the banks are completely vegetated. In contrast, the substrate in the Lakina, Gilahina, and Chokosna Rivers is gravel and large boulders. The banks of these rivers are nearly free of vegetation.

A total of 125 taxa were collected during the 2 years of sampling at the stream sites (appendix A). Of the insect taxa, 83 percent were insects, and of those 83 percent, 60 percent were flies and midges (Diptera). Mayflies (Ephemeroptera), stoneflies (Plecoptera), and caddisflies (Trichoptera) comprised 14, 12, and 11 percent of the total number of taxa, respectively; collectively, these are referred to as EPT taxa. Of the taxa collected and identified, 49 taxa had not been detected at 43 other sites in Alaska sampled using identical protocols. Of these 49 newly collected taxa, 37 were insects including 12 EPT taxa; however, most (23) were flies and midges. Only three taxa were common to all sites—blackflies (Diptera, *Simulium* sp.), water mites (Arachnida, *Sperchon* sp.), and a stonefly (*Zapada oregonensis*).

Using the taxonomic information from the macroinvertebrate samples and IDAS, eleven metrics were calculated (table 12). Metric values from the macroinvertebrate data (table 13) reveal some differences among sites, such as a relatively higher number of gatherer/collector taxa and lower number of EPT taxa at the outlet stream of Long Lake (15210300). This site is affected strongly by Long Lake and a lower stream gradient as compared to the Gilahina and Chokosna Rivers (15210600 and 15210700). As previously noted, glacially dominated streams tend to have harsh habitat conditions for aquatic biota and the Lakina River (15210200) QMH sample reflects that with relatively low richness. Other than those exceptions, macroinvertebrate communities had little variability among sites reflecting the undisturbed nature of the streams.

Table 11. Macrozooplankton data for Long Lake, Alaska.

[Values indicate number of individuals]

| Date | Macrozooplankton | | | | | | | | | | Macro-zooplankton total |
| | Small | | Large cladoceran | | Small copepod | Large copepod | Small copepod | Copepod | | | |
	Bosmina	Ceriodaphnia	Daphnia	Cyclopoid1	Cyclopoid2	Calanoid1	Calanoid2	Nauplii	Copepedid		
05-20-08	11	0	2	191	155	2	5	66	24		456
07-01-08	2	1	9	73	10	2	0	0	0		97
08-05-08	62	3	99	137	16	10	0	0	0		327
09-23-08	14	5	55	172	11	180	264	315	21		1,037

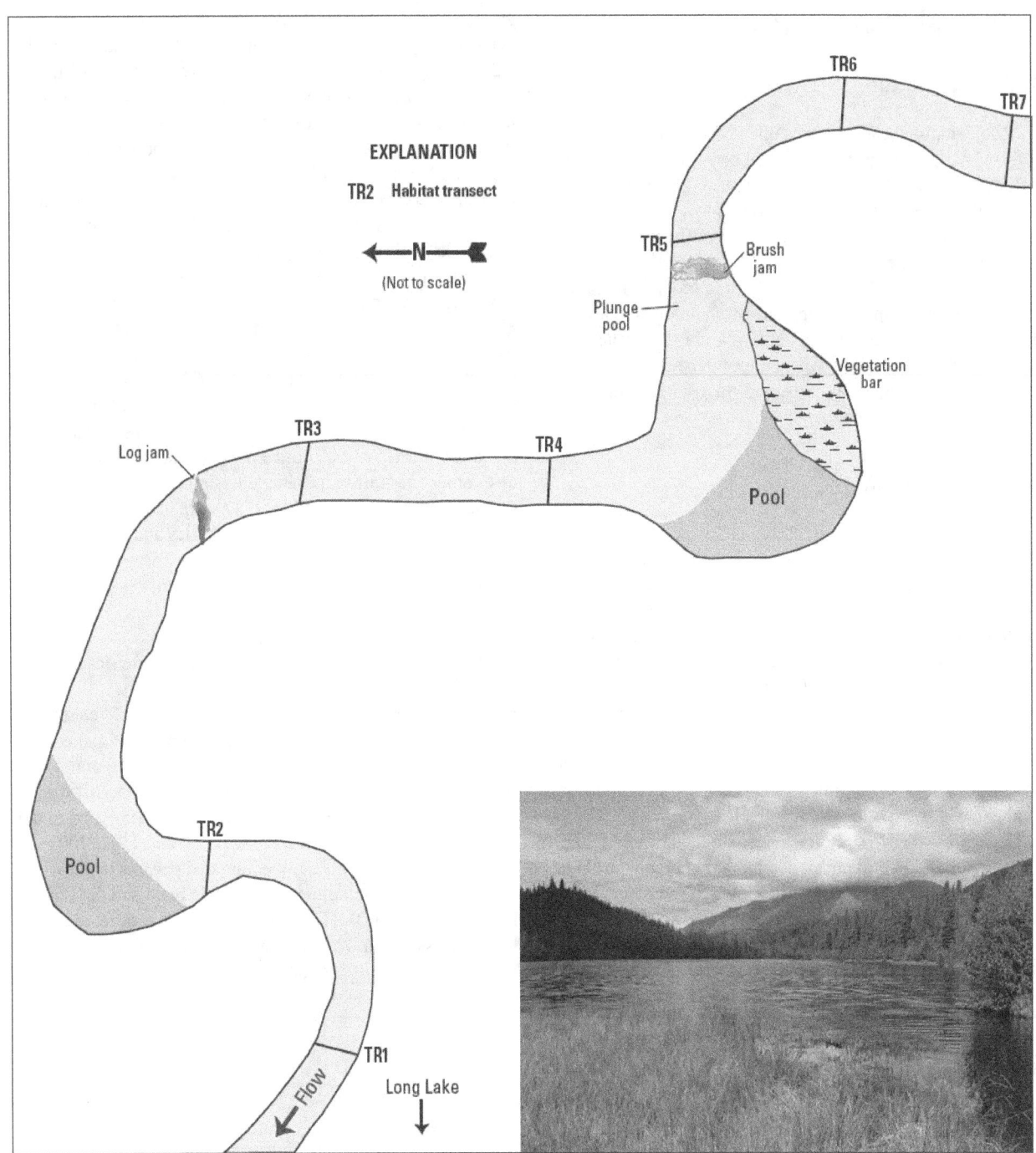

Figure 12.   Major geomorphic features and photograph of the Long Lake inlet stream study reach near McCarthy, Alaska.

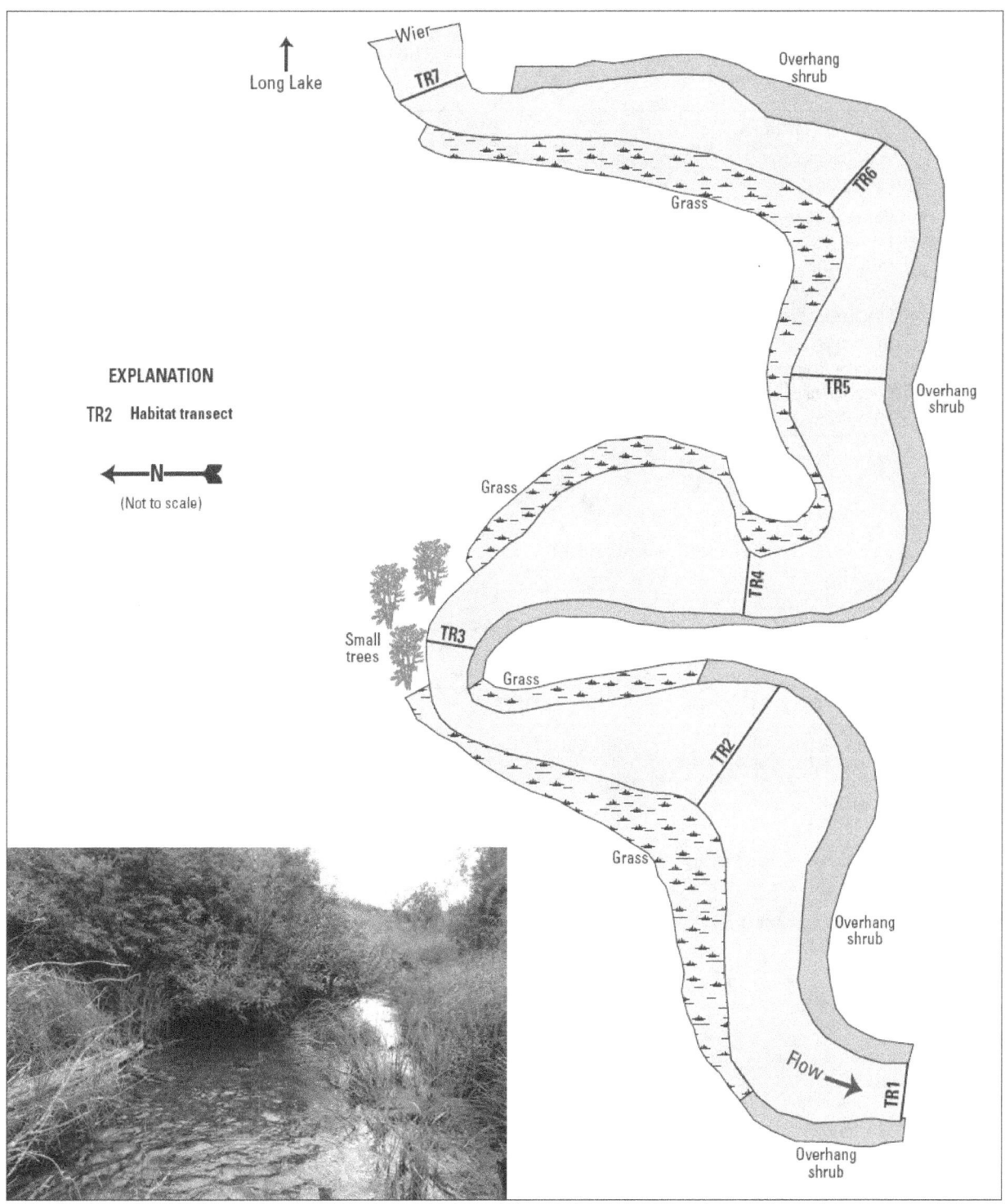

Figure 13. Major geomorphic features and photograph of the Long Lake outlet stream study reach, near McCarthy, Alaska.

Table 12. Definitions of macroinvertebrate metrics.

[**Abbreviations:** EPT, Ephemeroptera (mayfly), Plecoptera (stonefly), and Trichoptera (caddisfly); USEPA, U.S. Environmental Protection Agency]

Metric	Description
Taxa richness	Total richness (number of non-ambiguous taxa)
EPT richness	Richness composed of mayflies, stoneflies, and caddisflies
Percent EPT richness	Percentage of total richness composed of mayflies, stoneflies, and caddisflies
Percent chironomid richness	Percentage of total richness composed of midges
Richness of tolerant taxa	Average USEPA tolerance values for sample based on richness
Abundance of tolerant taxa	Abundance (density)-weighted USEPA tolerance value for sample
Shannon diversity	Shannon-Wiener diversity index
Evenness	Evenness (Shannon-Wiener diversity/maximum diversity)
Predator richness	Richness composed of predators
Collector/gatherer richness	Richness composed of collector-gatherers
V2DOM	Percentage of total abundance represented by the two most abundant taxon

Table 13. Metric values for all macroinvertebrate samples collected at sites along McCarthy Road, Alaska, 2007 and 2008.

[Metric definitions are shown in table 12. **Type:** Q, Qualitative Multi-Habitat; R, Richest Targeted Habitat. **Abbreviations:** EPT, Ephemeroptera (mayfly), Plecoptera (stonefly), and Trichoptera (caddisfly); V2DOM, Percentage of total abundance represented by the two most abundant taxon; NA, not applicable]

USGS station No.	Collection date	Type	Taxa richness	EPT richness	Percent EPT richness	Percent chironomid richness	Richness of tolerant taxa	Abundance of tolerant taxa	Shannon diversity	Evenness	Predator richness	Gatherer/collector richness	V2DOM
15210200	10-02-2007	Q	22	9	40.91	31.82	4.00	NA	NA	NA	4	9	NA
15210250	07-30-2007	Q	36	8	22.22	41.67	4.80	NA	NA	NA	7	13	NA
15210260	08-06-2008	R,Q	49	10	20.41	40.82	4 99	5.15	1.06	0.67	10	20	50.00
15210260	09-24-2008	R,Q	41	11	26.83	36.59	4.76	3.88	1.20	.78	8	14	26.56
15210300	07-29-2007	R	22	11	50.00	36.36	3.49	2.55	1.11	.83	4	7	37.55
15210300	07-31-2007	R,Q	55	10	18.18	41.82	5 21	4.67	1.16	.77	14	20	38.26
15210300	08-05-2008	R,Q	51	10	19.61	37.25	4 93	5.54	1.15	.73	10	17	44.86
15210600	08-01-2007	R,Q	42	14	33.33	40.48	4 29	2.01	.98	.68	9	10	43.50
15210600	09-25-2008	Q	34	13	38.24	26.47	3 94	NA	NA	NA	8	8	NA
15210700	07-29-2007	Q	30	10	33.33	43.33	4 35	NA	NA	NA	7	11	NA
15210700	09-25-2008	R,Q	32	14	43.75	21.88	3.89	3.96	1.10	.80	7	10	36.42

Another way to visually compare the macroinvertebrate communities among sites is through the use of an ordination resulting from a multivariate statistical analysis such as NMDS (figs. 14–15, table 14). For macroinvertebrate presence/absence data from sites sampled in 2007 and 2008 (fig. 14), the first (x) axis provides the greatest discrimination among sites and appears to distinguish the small, lower gradient, and wetland dominated sites on the left side of the diagram (Long Lake inlet and outlet streams) from the higher gradient sites or larger rivers on the right (Lakina, Gilahina, and Chokosna Rivers), supporting the interpretation of the metric data above. Sample M-3, the inlet stream to Long Lake, was collected at a relatively high discharge of 14 ft³/s, although the remaining samples at the site were collected at discharges less than 5 ft³/s, which may explain why it plotted between the two groups. The second (y) axis by definition has less discriminatory power than the first axis, and here may show the effect of the higher precipitation that occurred in 2008 relative to 2007. The 2008 samples generally plot higher that the 2007 samples with the exception of the 2008 sample

from the Long Lake outlet stream. As previously noted, this site is affected more by Long Lake than precipitation events. The Lakina River does not fall within an easily identified group, but rather plots far to the right indicating a high gradient. Not only is this a glacier-fed river, but also the largest basin with the highest flow at the time of sampling.

The second NMDS plot (fig. 15) shows that the McCarthy Road sites form a distinctly different group from the rest of the sites in Alaska sampled by the USGS using the same protocols. Of the 71 other sites (table 14), 34 are located on streams in Anchorage, and 21 streams are located on the Kenai Peninsula, south of Anchorage. These areas and the remaining sites represent a transitional climate between maritime and continental (Brabets and others, 1999). It is interesting to note that within the McCarthy Road sites, the larger, higher gradient sites form a fairly tight sub-group on the left side of the plot. Because most of the sites shown on this plot represent minimally disturbed areas, the distinction among groups likely reflects the effects of Interior Alaska climate.

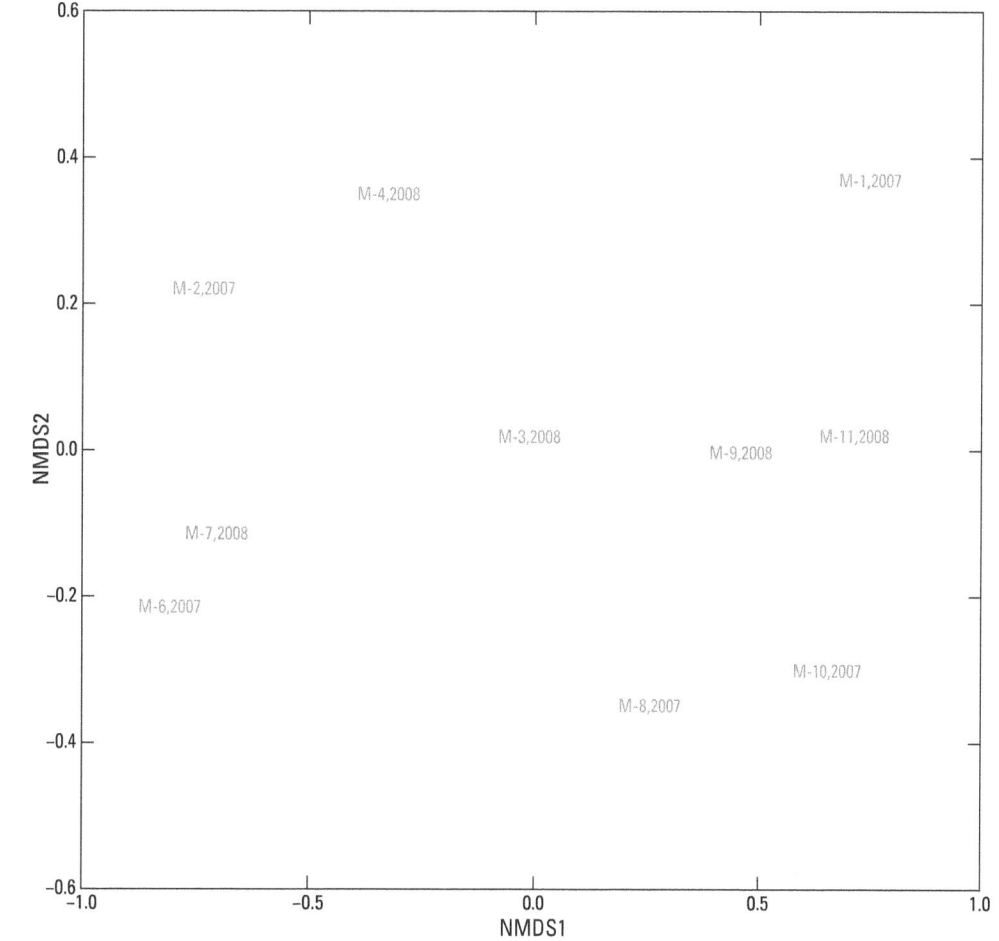

Figure 14. Similarity of macroinvertebrate communities at sites along McCarthy Road, Alaska, 2007 and 2008, based on the first and second dimension from a non-metric multi-dimensional scaling (NMDS) analysis.

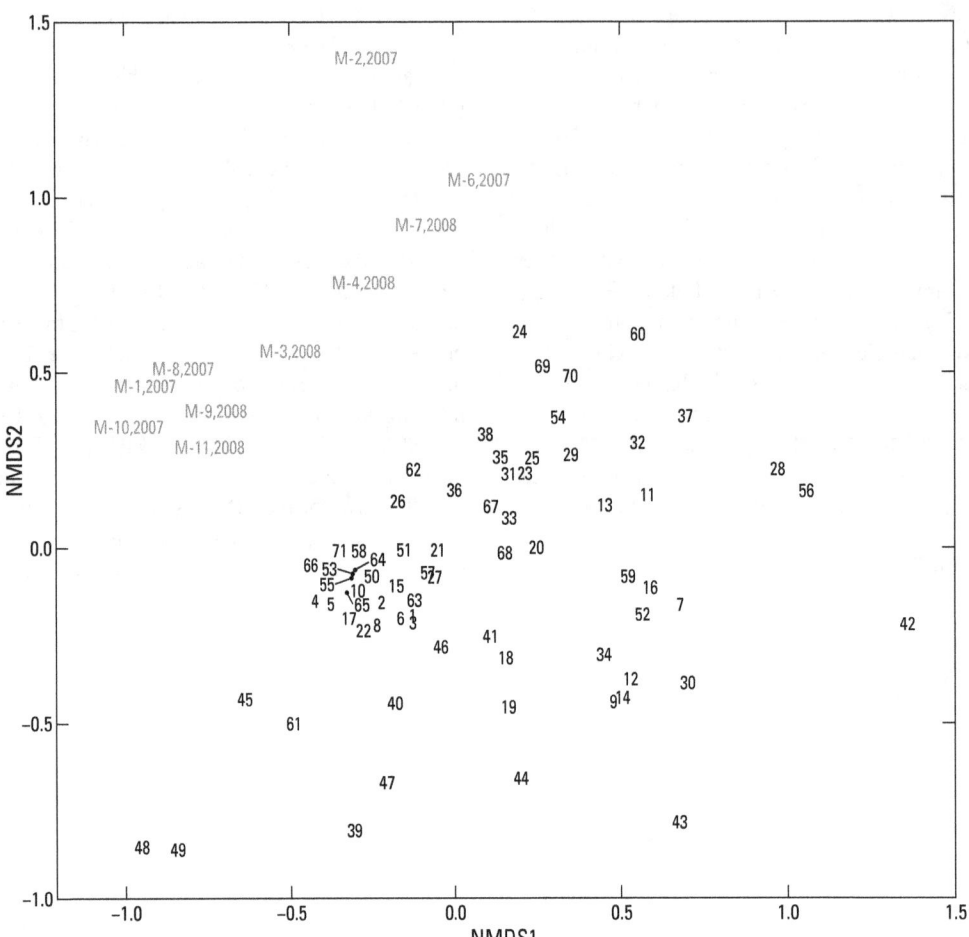

Figure 15.   Similarity of macroinvertebrate communities at sites along McCarthy Road, Alaska, 2007 and 2008, and at other sites in Alaska where USGS has collected macroinvertebrate data, based on the first and second dimension from a non-metric multi-dimensional scaling (NMDS) analysis.

Table 14. Stations used in non-metric multi-dimensional scaling analysis, Alaska.

Site No.	USGS station No.	Station name	Site No.	USGS station No.	Station name
1	15274000	South Fork Campbell Creek near Anchorage	41	601833154154100	Kijik River above Little Kijik River near Port Alsworth
2	15273040	Rabbit Creek at Porcupine Trail Road near Anchorage	42	601828154171700	Little Kijik River below Kijik Lake near Port Alsworth
3	15274000	South Fork Campbell Creek near Anchorage	43	15266020	Kenai River at Jims Landing near Cooper Landing
4	15273030	Rabbit Creek at East 140 Avenue near Anchorage	44	15266010	Kenai River below Russian River near Cooper Landing
5	15273020	Rabbit Creek at Hillside Drive near Anchorage	45	631018149323700	Costello Creek near Colorado
6	15274000	South Fork Campbell Creek near Anchorage	46	15292700	Talkeetna River near Talkeetna
7	15266110	Kenai River below Skilak Lake Outlet near Sterling	47	585750154101100	Kamishak River near Kamishak
8	15273100	Little Rabbit Creek near Anchorage	48	631629149352000	Colorado Creek near Colorado
9	15275100	Chester Creek at Arctic Boulevard at Anchorage	49	15294700	Johnson River above Lateral Glacier near Tuxedni Bay
10	15273090	Little Rabbit Creek at Nickleen Street near Anchorage	50	15273100	Little Rabbit Creek near Anchorage
11	15241600	Ninilchik River at Ninilchik	51	15274000	South Fork Campbell Creek near Anchorage
12	15275100	Chester Creek at Arctic Boulevard at Anchorage	52	15266110	Kenai River below Skilak Lake Outlet near Sterling
13	15294100	Deshka River near Willow	53	15273020	Rabbit Creek at Hillside Dr near Anchorage
14	15275100	Chester Creek at Arctic Boulevard at Anchorage	54	15241600	Ninilchik River at Ninilchik
15	15276200	Ship Creek at Glenn Hwy near Anchorage	55	15273040	Rabbit Creek at Porcupine Trail Road near Anchorage
16	15266300	Kenai River at Soldotna	56	15294100	Deshka River near Willow
17	15273097	Little Rabbit Creek at Goldenview Dr near Anchorage	57	15276200	Ship Creek at Glenn Highway near Anchorage
18	15274395	Campbell Creek at New Seward Highway near Anchorage	58	15273030	Rabbit Creek at East 140 Avenue near Anchorage
19	15274557	Campbell Creek at C Street near Anchorage	59	15266300	Kenai River at Soldotna
20	15274830	South Branch of South Fork Chester Creek at Boniface Parkway near Anchorage	60	15275100	Chester Creek at Arctic Boulevard at Anchorage
21	15276570	Ship Creek below Power Plant at Elmendorf Air Force Base	61	15292304	Costello Creek below Camp Creek near Colorado
22	15274796	South Branch of South Fork Chester Creek at Tank Trail near Anchorage	62	15292302	Camp Creek at Mouth near Colorado
23	15240300	Stariski Creek near Anchor Point	63	15283700	Moose Creek near Palmer
24	15240000	Anchor River at Anchor Point	64	15273090	Little Rabbit Creek at Nickleen Street near Anchorage
25	600204151401800	Deep Creek 0.6 mile above Sterling Highway near Ninilchik	65	15283550	Moose Creek above Wishbone Hill near Sutton
26	600107151112800	North Fork Deep Creek 4 miles above mouth near Ninilchik	66	15273097	Little Rabbit Creek at Goldenview Drive near Anchorage
27	15274000	South Fork Campbell Creek near Anchorage	67	15274395	Campbell Creek at New Seward Highway near Anchorage
28	15294100	Deshka River near Willow	68	15276570	Ship Creek below Power Plant at Elmendorf Air Force Base
29	600321151325000	Ninilchik River below Tributary 3 near Ninilchik	69	15274557	Campbell Creek at C Street near Anchorage
30	15266110	Kenai River below Skilak Lake Outlet near Sterling	70	15274830	South Branch of South Fork Chester Creek at Boniface Parkway near Anchorage
31	15241600	Ninilchik River at Ninilchik	71	15274796	South Branch of South Fork Chester Creek at Tank Trail near Anchorage
32	15275100	Chester Creek at Arctic Boulevard at Anchorage	M-1	15210200	Lakina River near McCarthy
33	594507151290000	Beaver Creek 2 miles above mouth near Bald Mountain near Homer	M-2	15210250	Long Lake Creek Tributary 2 at McCarthy Road near McCarthy
34	15266300	Kenai River at Soldotna	M-3	15210260	Long Lake Trib 1 150 Ft above Mouth near McCarthy
35	595506151403300	Stariski Creek 2 miles below Unnamed Tributary near Ninilchik	M-4	15210260	Long Lake Trib 1 150 Ft above Mouth near McCarthy
36	595126151391000	Chakok River 7.5 miles above mouth near Anchor Point	M-6	15210300	Long Lake Creek at McCarthy Road near McCarthy
37	600945151210900	Ninilchik River 1.5 miles below Tributary 1 near Ninilchik	M-7	15210300	Long Lake Creek at McCarthy Road near McCarthy
38	15239840	Anchor River above Twitter Creek near Homer	M-8	15210600	Gilahina River at McCarthy Road near Chitina
39	601708154203500	Little Kijik River above Kijik Lake near Port Alsworth	M-9	15210600	Gilahina River at McCarthy Road near Chitina
40	601801154143600	Kijik River 1.5 miles above Mouth near Port Alsworth	M-10	15210700	Chokosna River at McCarthy Road near Chitina
			M-11	15210700	Chokosna River at McCarthy Road near Chitina

Algae

Periphytic algal taxa richness and the number of unique taxa at each site (that is, the number of taxa at a given site in a given sample type—QMH or RTH—that were not at any other sites for a given sample type) were variable between sample type and sample locations (table 15, appendix B–C). No RTH sample was collected at the Lakina River due to the high flow and velocities. Taxa richness in RTH samples was consistently higher than taxa richness in QMH samples and the number of unique taxa in RTH samples was generally lower than the number of unique taxa in QMH samples; therefore, both QMH and RTH taxa richness and numbers of unique taxa are shown for a more complete representation of the algal community at a site. The Long Lake outflow sample collected on July 31, 2007, had the highest richness (59 taxa in the RTH sample and 52 in the QMH sample) and the highest number of unique taxa (19 in the RTH sample and 24 in the QMH sample). Generally, the Long Lake outflow had higher taxa richness and more unique taxa than the inflow tributaries, whereas the Long Lake inflow tributaries generally had higher taxa richness than the other river sites. At a given sample site, 60–100 percent of the taxa were non-motile, suggesting relatively stable habitat conditions.

The abundance of periphytic algal taxa identified to the five algae groups varied by sample location and by abundance measure (density compared with biovolume) (fig. 16). Algae groups are defined by phylum, with the exception of diatoms and yellow-green algae, which are defined by class (appendix C). Algae groups reported here, including green algae (Chlorophyta), Diatoms (Bacillariophyceae), Yellow-Green Algae (Chrysophyceae), Blue-Green Algae (Cyanophyta), and Red Algae (Rhodophyta) are all taxa commonly found in lotic (swift moving water) ecosystems. The taxa with the highest relative abundance, as measured by density and biomass, in the inflow tributaries to Long Lake were the red algae (47–96 percent), a community mainly of an unknown rhodophyte (appendix C). In contrast, blue-green algae had the highest relative abundance (about 50 percent), as measured by density, at the Long Lake outflow. The remaining 50 percent of the taxa at the Long Lake outflow, as measured by density, were green algae, diatoms, and red algae. However, 92 percent of the relative abundance, as measured by biovolume, in samples collected from Long Lake outflow on August 5, 2008, were green algae that are composed mainly of the large-bodied chlorophyte (*Spirogyra* sp.), a filamentous algae common in various freshwater environments. Spirogyra often are free floating, which may contribute to their high biovolume at the outlet to Long Lake. At the Gilahina and Chokosna River sites, the type of

Table 15.   Periphytic algae taxa richness for Qualitative Multi-Habitat and Richest Targeted Habitat samples and Simpson index of diversity values for Richest Targeted Habitat samples, and number of unique taxa at sites along McCarthy Road, Alaska, 2007 and 2008.

[**Abbreviations:** QMH, Qualitative Multi-Habitat; RTH, Richest Targeted Habitat; –, sample not collected]

Date	Taxa richness		Simpson diversity	Number of unique taxa	
	QMH sample	RTH sample		QMH sample	RTH sample
Lakina River					
10-02-07	21	–	–	4	–
Long Lake Inflow-Tributary 1					
08-06-08	19	44	0.33	8	6
09-24-08	–	32	.72	–	3
Long Lake Inflow-Tributary 2					
07-30-07	40	48	0.27	12	13
Long Lake Outflow					
07-31-07	52	59	0.84	24	19
08-05-08	25	49	.85	8	13
Gilahina River					
08-01-07	19	33	0.57	5	1
Chokosna River					
07-29-07	9	14	0.49	2	1
08-04-08	–	29	.85	–	12

abundance measure used (density compared with biovolume) produced similar trends in the relative abundance data. The Gilahina River, as measured by a density to biovolume ratio, was dominated by two cyanophytes, *Homoeothrix janthina* and one unknown bloom. The yellow-green algae *Hydrurus foetidus* was the dominant taxa in the Chokosna River in 2007, yet there was no yellow-green algae detected in the 2008 sample. The dominant taxa, with respect to density, were the unknown cyanophyte and *Chamaesiphon* sp. (also a cyanophyte), as well as the undetermined *Centric* sp. diatom. With respect to biovolume, the undetermined *Centric* sp. diatom was the most dominant taxa, followed by *Hannaea arcus* and the cyanophyte *Chamaesiphon* sp., respectively, in the Chokosna River in 2008. Accurate assessment of specific drivers (for example, physical habitat) of temporal change in algal community composition requires taxa-specific habitat preference information and is beyond the scope of this study.

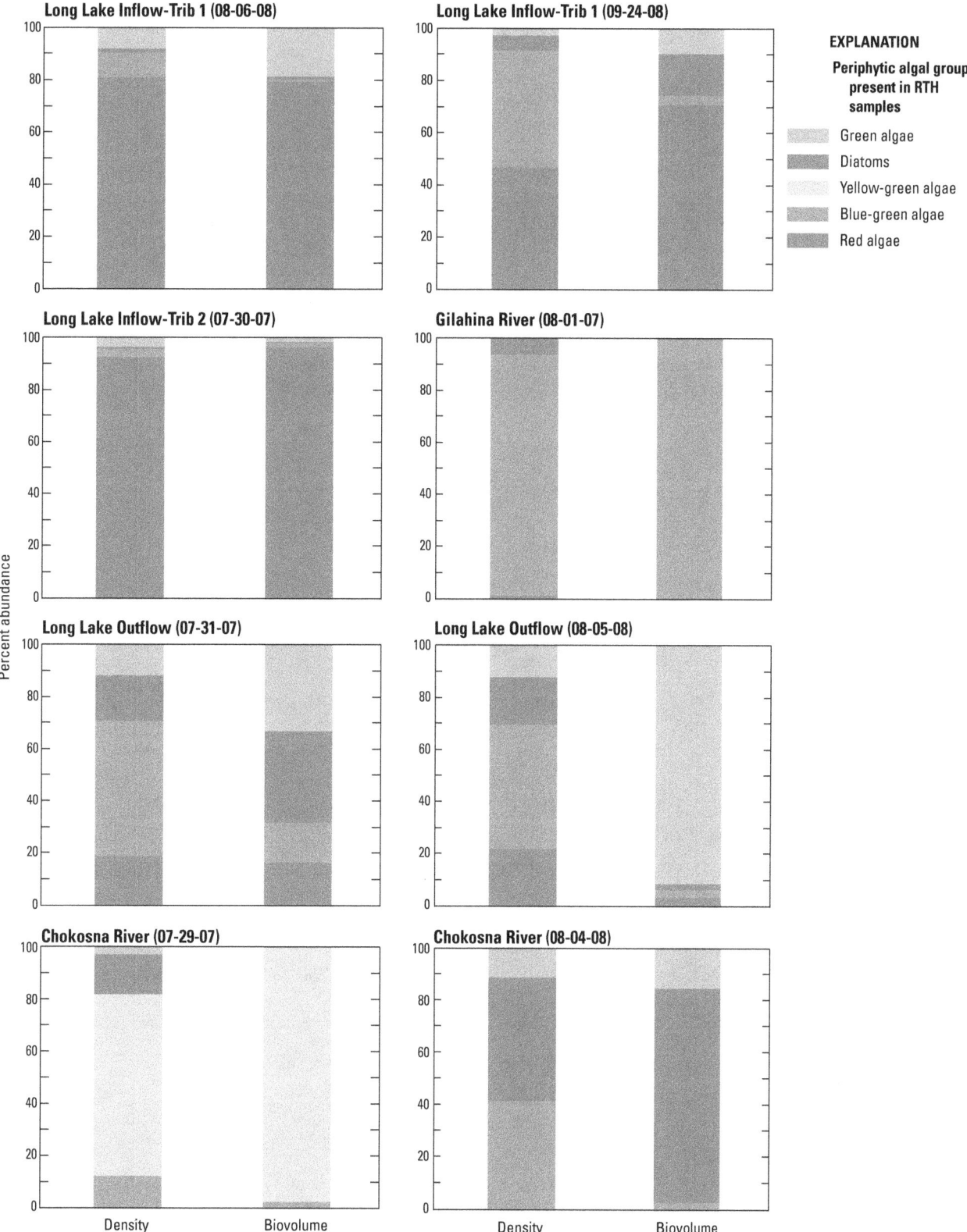

Figure 16. Percent abundance as density and biovolume of periphytic algal groups in richest targeted habitat (RTH) samples collected at sites along McCarthy Road, Alaska, 2007 and 2008.

However, it is possible that difference between years may be a result of difference in physical habitat conditions (for example, higher precipitation and higher discharge in 2008).

In addition to taxa richness and abundance, diversity is a common metric used to assess community structure. Diversity measures integrate information on both taxa richness and evenness (relative abundance of each taxon). Values of the Simpson index of diversity range from 0 to 1, with larger numbers indicating greater diversity. The Simpson index of diversity at the Long Lake inflow sites was variable with values ranging from 0.27 to 0.72 (table 15). In contrast, diversity was consistently high at the Long Lake outlet (0.85). The Simpson index of diversity was intermediate at the Gilahina River site (0.57) and varied from 0.49 to 0.85 at the Chokosna River. Temporal and spatial differences in algal diversity may be reflective of changing biotic or abiotic conditions, including environmental and (or) physical habitat conditions. Our ability to identify which biotic or abiotic

variables are important drivers of change in algal diversity is limited by data availability. However, the diversity values presented in table 15 may be informative to future studies at these sites. For example, differences in biogeochemical nutrient cycles between sites may be influenced by the abundance and diversity of the algae at a given site.

A NMDS ordination plot of the eight RTH samples is shown in figure 17. The two-dimensional stress associated with the ordination was 0.05, indicating that the plot gives an excellent representation of the relations between samples. The two Long Lake outflow samples grouped close together and had periphytic algal communities that were 41 percent similar. Similarly, the Long Lake inflow samples and the Gilahina River samples were grouped together and had communities that were 30–58 percent similar. On the other hand, the two Chokosna River samples had unique algal communities that were not similar between years or to any of the other samples.

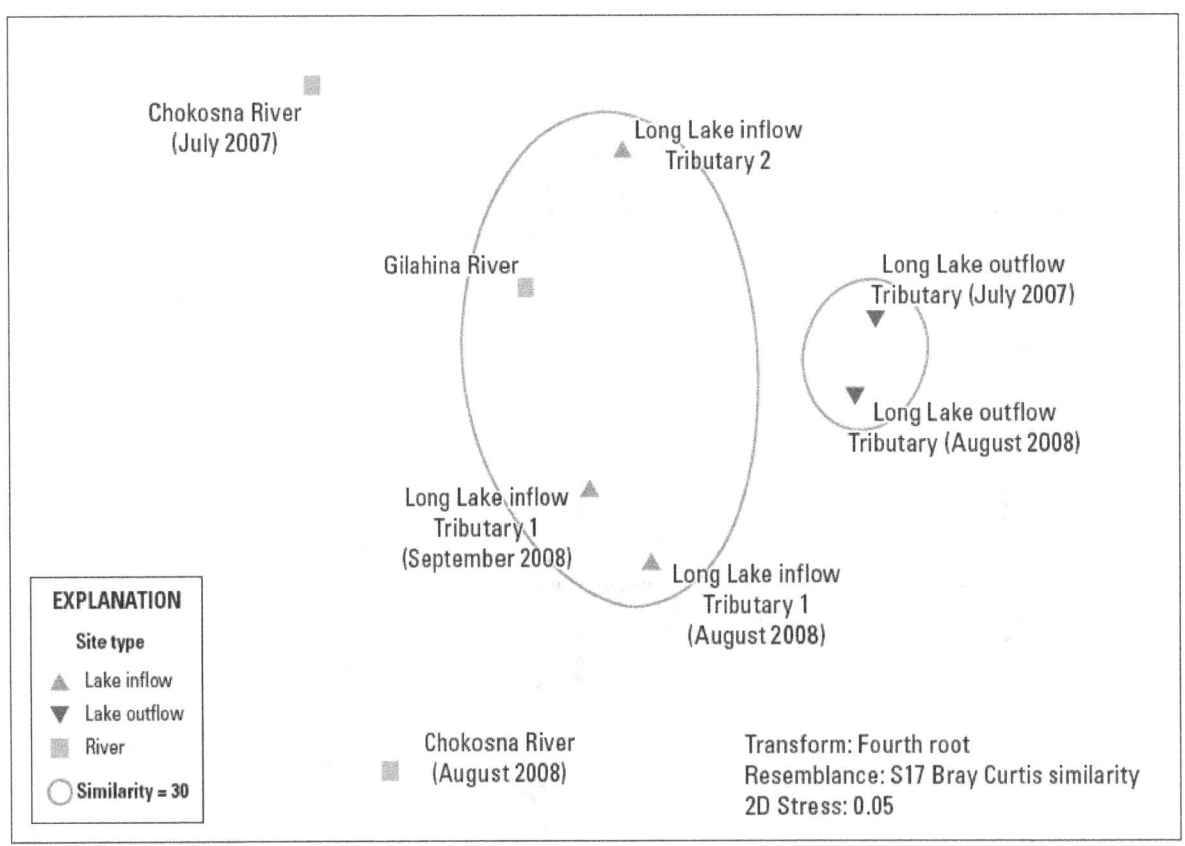

Figure 17.   Non-metric multi-dimensional scaling ordination plot of richest targeted habitat periphytic algae biovolume data from samples collected at sites along McCarthy Road, Alaska, 2007 and 2008.

Summary

The Chokosna, Gilahina, and Lakina Rivers, and Long Lake watershed, along McCarthy Road within Wrangell-St. Elias National Park and Preserve (WRST), Alaska, support popular recreational activities as well as an important commercial sockeye salmon fishery of the Copper River. To gain a better knowledge of the hydrology and limnology of this area, a study of the streams and the Long Lake system was undertaken in 2007 and 2008. Major findings of the study include:

- Water samples from all sites indicate the water type at all sites is calcium bicarbonate—with the exception of the Chokosna River, which is calcium bicarbonate sulfate. Alkalinity concentrations ranged from 63 to 222 milligrams per liter, indicating a high buffering capacity of these waters.

- During the 2 years of data collection recorded water temperatures did not exceed 10 °C at the Chokosna and Gilahina Rivers. At the Lakina River, a glacier-fed river, water temperatures were less than 9 °C. The most pronounced differences in water temperature were noted in the inlet and outlet streams of Long Lake. Water temperatures for the inlet stream were as high as 15 °C and as high as 19 °C at the outlet stream.

- A water temperature logistic model was calibrated with known air temperature and water temperature from 2007 and 2008 and then used to simulate water temperature from 1998 to 2006 for the Long Lake outlet stream. Analysis of the stream temperature during this period showed no trends but did indicate the 2007 water temperatures were the highest during this period.

- Depth profiles of Long Lake show strong temperature stratification during the summer from the surface to about 13 feet with water temperatures ranging from 16 to 5 °C. Dissolved oxygen profiles of Long Lake show strong stratifications between 26 and 33 feet, below which the concentrations of dissolved oxygen decreases from 10 to 2 mg/L.

- Analyses of trace elements in samples of streambed sediments collected from five sites indicated that concentrations of arsenic, chromium, and nickel exceeded levels that are believed to cause adverse effects to aquatic habitat. Given the nearly pristine nature of the study area, however, these concentrations most likely reflect local geology of the area rather than anthropogenic sources.

- One hundred twenty-five macroinvertebrate taxa were identified over the 2-year sampling period. Eighty-three percent of macroinvertebrates were insects. Dipterans (flies and midges) accounted for 43 percent of all aquatic macroinvertebrates found at the 6 sites. Forty-nine new taxa, not found in other USGS sampled sites using NAWQA protocols, were identified in the samples collected along McCarthy Road.

- Analysis of the macroinvertebrate data by non-metric multi-dimensional scaling (NMDS) showed a separation of sites by size of the stream gradient or proximity to Long Lake and a separation between the 2007 and 2008 samples, likely due to the higher precipitation in 2008. A second NMDS of the McCarthy Road macroinvertebrate data with other Alaska macroinvertebrate data collected by the USGS showed a distinct separation between the two data sets. This separation suggests that macroinvertebrate data at sites along the McCarthy Road represent a continental type climate; other USGS Alaska macroinvertebrate data represent a transitional climate between maritime and continental.

- Periphytic algal taxa richness, abundance, and community composition was variable between sites. Taxa richness was highest at the Long Lake outflow, suggesting that the lake may have contributed planktonic taxa to the periphytic community and (or) created physical and chemical conditions at the outlet favorable to a variety of taxa. Lake inflow and river sites tended to be dominated by specific taxa. Specifically, red algae were the dominant taxa at the lake inflow sites and blue-green algae were abundant at the Gilahina River. Shifts in the dominant taxa between years at the Chokosna River may be indicative of differences in physical and chemical conditions between the two sample times.

- Site groupings defined by NMDS of periphytic algal communities suggest that the Long Lake outflow site formed one group and Long Lake inflow tributaries and the Gilahina River formed another group. However, periphytic algal communities in the Chokosna River samples were not similar to each other between years to any of the other samples.

Acknowledgments

The authors wish to thank Eric Veach of Wrangell-St. Elias National Park and Preserve for his support of this study.

References Cited

Aiken, G.R., and Cotsaris, E., 1995, Soil and Hydrology–their effect on NOM: American Water Works Association, January, p. 26-45.

Alaska National Interest Lands Conservation Act (ANILCA), 1980, 16 U.S.C. 3101 et seq (1988), Dec. 2 1980, Stat. 2371, Pub. L. 96-487. Alaska Native Claims Settlement Act (ANCSA). Dec. 18, 1971, Pub. L. 92-203.

Arbogast, B.F., ed., 1990, Quality assurance manual for the Branch of Geochemistry Survey: U.S. Geological Survey Open-File Report 90-668, 184 p.

Balcer, B.D., Korda, N.L., and Dodson, S.I., 1984, Zooplankton of the Great Lakes: London, England, The University of Wisconsin Press, Ltd. p. 56-58.

Brabets, T.P., Nelson, G.L., Dorava, J.M., and Milner, A.M., 1999, Water-quality assessment of the Cook Inlet Basin, Alaska–Environmental Setting: U.S. Geological Survey Water-Resources Investigations Report 99-4025, 65 p.

Brabets, T.P., and Whitman, M.S., 2002, Water quality of Camp Creek, Costello Creek, and other selected streams on the south side of Denali National Park and Preserve, Alaska: U.S. Geological Survey Water-Resources Investigations Report 02-4260, 52 p.

Canadian Council of Ministers of the Environment, 1999, Canadian sediment quality guidelines for the protection of aquatic life–Summary tables, in Canadian Environmental Quality Guidelines, 1999: Winnipeg, Canada, Canadian Council of Ministers of the Environment.

Clarke, K.R., and Gorley, R.N., 2006, PRIMER v6–User Manual/Tutorial: Plymouth, England, PRIMER-E.

Clarke, K.R., Somerfield, P.J., and Gorley, R.N., 2008, Testing of null hypotheses in exploratory community analyses–Similarity profiles and biota-environment linkage: Journal of Experimental Marine Biology and Ecology, v. 336, p. 56-69.

Clarke, K.R., and Warwick, R.M., 2001, Change in marine communities–An approach to statistical analysis and interpretation, 2nd edition: Plymouth, England, PRIMER-E ltd., 168 p.

Cuffney, T.F., and Brightbill, R.A., 2011, User's manual for the National Water-Quality Assessment Program Invertebrate Data Analysis System (IDAS) software, version 5: U.S. Geological Survey Techniques and Methods 7-C4, 126 p. (Available at http://pubs.usgs.gov/tm/7c4/.)

Cuffney, T.F., Gurtz, M.E., and Meador, M. R., 1993, Methods for collecting benthic invertebrate samples as part of the National Water-Quality Assessment Program: U.S. Geological Survey Open-File Report 93-406, 66 p.

Deacon, J.R., and Stephens, V.C., 1998, Trace elements in streambed sediment and fish liver at selected sites in the Upper Colorado River Basin, Colorado, 1995–96: U.S. Geological Survey Water-Resources Investigations Report 98-4124, 19 p.

Decho, A.W., and Luoma, S.N., 1994, Humic and fulvic acids-sink or source in the availability of metals to the marine bivalves Macoma balthica and Potamocorbula amurensis; Marine Ecology Progress Series, v. 108, p. 133-145.

Fishman, M.J., 1993, Methods of analysis by the U.S. Geological Survey National Water-Quality Laboratory–Determination of inorganic and organic constituents in water and fluvial sediments: U.S. Geological Survey Open-File Report 93-125, 217 p.

Fishman, M.J., and Friedman, L.C., eds., 1989, Method for determination of inorganic substances in water and fluvial sediments: U.S. Geological Survey Techniques of Water-Resources Investigations, Book 5, Chapter A1, 545 p.

Gaudy, A.F., and Gaudy, E.T., 1988, Elements of bioenvironmental engineering: San Jose, Calif., Engineering Press, Inc., 592 p.

Gibbs, R.J., 1973, Mechanisms of trace metal transport in rivers: Science, v. 180, p. 71-73.

Gislason, G.M., Olafsson, J.S., and Adalsteinsson, H., 2000, Life in glacial and alpine rivers in central Iceland in relation to physical and chemical parameters: Nordic Hydrology, v. 31, no. 4–5, p. 411-422.

Hem, J.D., 1985, Study and interpretation of the chemical characteristics of natural water: U.S. Geological Survey Water-Supply Paper 2254, 264 p.

Horne, A.J., and Goldman, C.R., 1994, Limnology: New York, McGraw-Hill, Inc., 576 p.

Horowitz, A.J., 1991, A primer on sediment-trace element chemistry (2nd ed.): Chelsea, Mich., Lewis Publishers, 136 p.

Horowitz, A.J., and Stephens, V.C., 2008, The effects of land use on fluvial sediment chemistry for the conterminous U.S., results from the first cycle of the NAWQA Program–Trace and major Elements, phosphorus, carbon, and sulfur: Science of the Total Environment, v. 400, no. 1–3, doi:10.1016/j.scitoenv.2008.04.027, p. 290 -314.

Kyle, R.E., and Brabets, T. P., 2001, Water temperature of streams in the Cook Inlet Basin, Alaska, and implications of climate change: U.S. Geological Survey Water-Resources Investigations Report 01-4109, 24 p.

MacDonald, D.D., Ingersoll, C.G., and Berger, T.A., 2000, Development and evaluation of consensus-based sediment quality guidelines for freshwater ecosystems: Archives of Environmental Contamination and Toxicology, v. 39, p. 20-31.

Mohseni, O., Stefan, H.G., and Erickson, T.R., 1998, A nonlinear regression model for weekly stream temperature: Water Resources Research, v. 34, no. 10, p. 2685-2692.

Nash, J.E., and Sutcliffe, J.V., 1970, River flow forecasting through conceptual models: Journal of Hydrology, v. 10, p. 282-290.

National Park Service, 1998, Resources management plan—Wrangell-St. Elias National Park and Preserve, Copper Center, Alaska: National Park Service.

Newman, M.C., and Jagoe, C.H., 1994, Ligands and the bioavailability of metals in aquatic environments, *in* Hamelink, J.L., Landrum, P.F., Bergman, H.L., and Benson, W.H., eds., Bioavailability-physical, chemical, and biological interactions: Boca Raton, Fla., Lewis Publishers/ CRC Press, p. 39-61.

Oksanen, J. F., Blanchet, G., Kindt, R., Legendre, P., O'Hara, R. B,. Simpson, G.L., Solymos, P., Stevens, M.H.H., and Wagner, H., 2010, Vegan—Community Ecology Package: Free Software Foundation, Inc., R package version 1.17-5, accessed July 13, 2011, at http://CRAN.R-project.org/ package=vegan.

Patton, C.J., and Truitt, E.P., 1992, Methods of analysis by the U.S. Geological Survey National Water-Quality Laboratory—Determination of total phosphorus by a Kjeldahl digestion method and an automated colorimetric finish that includes dialysis: U.S. Geological Survey Open-File Report 92-146, 39 p.

Porter, S.D., Cuffney, T.F., Gurtz, M.E., and Meador, M.R., 1993, Methods for collecting algal samples as part of the National Water-Quality Assessment Program: U.S. Geological Survey Open-File Report 93-409, 39 p.

R Development Core Team, 2010, R–A language and environment for statistical computing: Vienna, Austria, R Foundation for Statistical Computing, Version 2.11.1, accessed July 13, 2011, at http://www.R-project.org.

Reiser, D.W., and Bjornn, T.C., 1979, Habitat requirements of anadromous salmonids: U.S. Forest Service, Pacific Northwest Research Station, General Technical Report PNW-96, 54 p.

Richter, D.H., Preller, C.C., Labay, K.A., and Shew, N.B., 2006, Geologic map of the Wrangell-St Elias National Park and Preserve, Alaska: U.S. Geological Survey Scientific Investigations Map 2877, 14 p. 1 pl.

Shelton, L.R., 1994, Field guide for collecting and processing stream-water samples for the National Water-Quality Assessment Program: U.S. Geological Survey Open-File Report 94-455, 42 p.

Shelton, L.R., and Capel, P.D., 1994, Guidelines for collecting and processing samples of stream bed for analysis of trace elements and organic contaminants for the National Water-Quality Assessment Program: U.S. Geological Survey Open-File Report 94-458, 20 p.

Shulski, M., and Wendler, G., 2007, The climate of Alaska: Fairbanks, Alaska, University of Alaska Press, 214 p.

Spacie, A., and Hamelink, J.L., 1985, Bioaccumulation, chapter 17 in Fundamentals of aquatic toxicology: New York, Hemisphere Publishing Corporation, p. 495-525.

Turnipseed, D.P., and Sauer, V.B., 2010, Discharge measurements at gaging stations: U.S. Geological Survey Techniques and Methods book 3, chap. A8, 87 p. (Also available at http://pubs.usgs.gov/tm/tm3-a8/.)

U.S. Environmental Protection Agency, 1976, Quality criteria for water: U.S. Environmental Protection Agency, Washington, D.C., 256 p.

U.S. Geological Survey, variously dated, National field manual for the collection of water-quality data: U.S. Geological Survey Techniques of Water-Resources Investigations, book 9, chaps. A1-A9. (Also available at http://pubs.water.usgs.gov/twri9A/).

Van Derveer, W.D., and Canton, S., 1997, Selenium sediment toxicity thresholds and derivation of water quality criteria for freshwater biota of western streams: Environmental Toxicology and Chemistry, v. 16, p. 1,260-1,268.

Venables, W.N., and Ripley, B.D., 2002, Modern applied statistics with S (3d ed.): New York, Springer, 501 p.

Wagner, R.J., Boulger, R.W., Jr., Oblinger, C.J., and Smith, B.A., 2006, Guidelines and standard procedures for continuous water-quality monitors—Station operation, record computation, and data reporting: U.S. Geological Survey Techniques and Methods Report 1-D3, 96 p.

Ward, H.B., Whipple, G.C. and Edmonson, W.T., eds., 1959, Fresh water biology (2nd ed.): New York, John Wiley & Sons, Inc., p. 1,248.

Winner, R.W., 1985, Bioaccumulation and toxicity of copper as affected by interactions between humic acid and water hardness: Water Resources Research, v. 19, no. 4, p. 449-455.

Appendix A. Benthic Macroinvertebrate Community Detected at Stream Sites along McCarthy Road, Wrangell-St. Elias National Park and Preserve, Alaska, 2007–08

Table A1. Benthic macroinvertebrate community detected at stream sites along McCarthy Road, Wrangell-St. Elias National Park and Preserve, Alaska, 2007–08.

Count	Phylum	Class	Order	SubOrder	Family	SubFamily	Tribe	Genus	Species
1	Cnidaria	Hydrozoa	Hydroida		Hydridae			Hydra sp.	
2	Platyhelminthes	Turbellaria	Tricladida		Planariidae			Polycelis sp.	
3	Platyhelminthes	Turbellaria							
4	Nematoda								
5	Mollusca	Gastropoda	Basommatophora		Lymnaeidae			Lymnaea sp.	
6	Mollusca	Gastropoda	Basommatophora		Lymnaeidae				
7	Mollusca	Gastropoda	Basommatophora		Planorbidae			Gyraulus sp.	
8	Mollusca	Gastropoda	Mesogastropoda		Valvatidae			Valvata sp.	Valvata lewisi
9	Mollusca	Gastropoda	Mesogastropoda		Valvatidae			Valvata sp.	
10	Mollusca	Bivalvia	Veneroida		Sphaeriidae			Pisidium sp.	
11	Mollusca	Bivalvia	Veneroida		Sphaeriidae	Pisidiinae			
12	Annelida	Oligochaeta							
13	Annelida	Hirudinea	Rhynchobdellae		Glossiphoniidae				
14	Arthropoda	Arachnida	Prostigmata		Lebertiidae			Lebertia sp.	
15	Arthropoda	Arachnida	Prostigmata		Sperchontidae			Sperchon sp.	
16	Arthropoda	Arachnida	Prostigmata		Sperchontidae			Sperchonopsis sp.	
17	Arthropoda	Arachnida	Sarcoptiformes	Oribatei					
18	Arthropoda	Arachnida	Trombidiformes	Prostigmata	Hygrobatidae			Atractides sp.	
19	Arthropoda	Arachnida	Trombidiformes	Prostigmata	Hygrobatidae			Hygrobates sp.	
20	Arthropoda	Ostracoda							
21	Arthropoda	Malacostraca	Amphipoda	Gammaridea	Gammaridae			Gammarus sp.	
22	Arthropoda	Malacostraca	Amphipoda	Gammaridea	Hyalellidae			Hyalella sp.	
23	Arthropoda	Insecta	Collembola		Isotomidae				
24	Arthropoda	Insecta	Collembola		Sminthuridae				
25	Arthropoda	Insecta	Ephemeroptera	Furcatergalia	Caenidae			Caenis sp.	
26	Arthropoda	Insecta	Ephemeroptera	Furcatergalia	Ephemerellidae			Drunella sp.	Drunella doddsi
27	Arthropoda	Insecta	Ephemeroptera	Furcatergalia	Leptohyphidae			Tricorythodes sp.	
28	Arthropoda	Insecta	Ephemeroptera	Furcatergalia	Leptophlebiidae				
29	Arthropoda	Insecta	Ephemeroptera	Furcatergalia	Leptophlebiidae			Paraleptophlebia sp.	
30	Arthropoda	Insecta	Ephemeroptera	Pisciforma	Ameletidae			Ameletus sp.	
31	Arthropoda	Insecta	Ephemeroptera	Pisciforma	Baetidae			Baetis sp.	
32	Arthropoda	Insecta	Ephemeroptera	Pisciforma	Baetidae			Baetis sp.	Baetis bicaudatus
33	Arthropoda	Insecta	Ephemeroptera	Pisciforma	Baetidae			Baetis sp.	Baetis flavistriga
34	Arthropoda	Insecta	Ephemeroptera	Pisciforma	Baetidae			Baetis sp.	Baetis tricaudatus
35	Arthropoda	Insecta	Ephemeroptera	Pisciforma	Baetidae			Procloeon sp.	
36	Arthropoda	Insecta	Ephemeroptera	Setisura	Heptageniidae			Cinygmula sp.	
37	Arthropoda	Insecta	Ephemeroptera	Setisura	Heptageniidae			Epeorus sp.	
38	Arthropoda	Insecta	Ephemeroptera	Setisura	Heptageniidae			Epeorus sp.	Epeorus deceptivus
39	Arthropoda	Insecta	Ephemeroptera	Setisura	Heptageniidae			Epeorus sp.	Epeorus grandis
40	Arthropoda	Insecta	Plecoptera	Euholognatha	Capniidae				

Table A1.   Benthic macroinvertebrate community detected at stream sites along McCarthy Road, Wrangell-St. Elias National Park and Preserve, Alaska, 2007–08.—Continued

Count	Phylum	Class	Order	SubOrder	Family	SubFamily	Tribe	Genus	Species
36	Arthropoda	Insecta	Plecoptera	Euholognatha	Capniidae	Capniinae		*Eucapnopsis* sp.	*Eucapnopsis brevicauda*
37	Arthropoda	Insecta	Plecoptera	Euholognatha	Nemouridae	Amphinemurinae		*Amphinemura* sp.	
38	Arthropoda	Insecta	Plecoptera	Euholognatha	Nemouridae	Nemourinae		*Nemoura* sp.	
39	Arthropoda	Insecta	Plecoptera	Euholognatha	Nemouridae	Nemourinae		*Zapada* sp.	
40	Arthropoda	Insecta	Plecoptera	Euholognatha	Nemouridae	Nemourinae		*Zapada* sp.	*Zapada cinctipes*
41	Arthropoda	Insecta	Plecoptera	Euholognatha	Nemouridae	Nemourinae		*Zapada* sp.	*Zapada oregonensis* group
42	Arthropoda	Insecta	Plecoptera	Euholognatha	Taeniopterygidae	Brachypteryinae		*Taenionema* sp.	
43	Arthropoda	Insecta	Plecoptera	Systellognatha	Chloroperlidae	Chloroperlinae		*Suwallia* sp.	
44	Arthropoda	Insecta	Plecoptera	Systellognatha	Perlodidae	Isoperlinae		*Isoperla* sp.	
45	Arthropoda	Insecta	Plecoptera	Systellognatha	Pteronarcyidae	Pteronarcyinae	Pteronarcellini	*Pteronarcella* sp.	
46	Arthropoda	Insecta	Trichoptera	Integripalpia	Brachycentridae	Brachycentrinae		*Brachycentrus* sp.	*Brachycentrus americanus*
47	Arthropoda	Insecta	Trichoptera	Integripalpia	Brachycentridae			*Micrasema* sp.	
48	Arthropoda	Insecta	Trichoptera	Integripalpia	Limnephilidae	Dicosmoecinae		*Ecclisomyia* sp.	
49	Arthropoda	Insecta	Trichoptera	Integripalpia	Limnephilidae	Dicosmoecinae		*Onocosmoecus* sp.	
50	Arthropoda	Insecta	Trichoptera	Integripalpia	Limnephilidae	Limnephilinae		*Limnephilus* sp.	
51	Arthropoda	Insecta	Trichoptera	Integripalpia	Limnephilidae	Limnephilinae		*Psychoglypha* sp.	
52	Arthropoda	Insecta	Trichoptera	Integripalpia	Limnephilidae	Limnephilinae	Chilostigmini	*Glyphopsyche* sp.	
53	Arthropoda	Insecta	Trichoptera	Spicipalpia	Glossosomatidae	Glossosomatinae		*Glossosoma* sp.	
54	Arthropoda	Insecta	Trichoptera	Spicipalpia	Hydroptilidae	Hydroptilinae		*Hydroptila* sp.	
55	Arthropoda	Insecta	Trichoptera	Spicipalpia	Hydroptilidae	Hydroptilinae		*Ochrotrichia* sp.	
56	Arthropoda	Insecta	Trichoptera	Spicipalpia	Hydroptilidae	Hydroptilinae		*Oxyethira* sp.	
57	Arthropoda	Insecta	Trichoptera	Spicipalpia	Rhyacophilidae			*Rhyacophila* sp.	*Rhyacophila vofixa* group
58	Arthropoda	Insecta	Lepidoptera						
59	Arthropoda	Insecta	Diptera	Nematocera	Ceratopogonidae	Ceratopogoninae		*Bezzia/Palpomyia* sp.	
60	Arthropoda	Insecta	Diptera	Nematocera	Ceratopogonidae	Ceratopogoninae		*Probezzia* sp.	
61	Arthropoda	Insecta	Diptera	Nematocera	Chironomidae	Chironominae	Chironomini	*Cryptochironomus* sp.	
62	Arthropoda	Insecta	Diptera	Nematocera	Chironomidae	Chironominae	Chironomini	*Microtendipes* sp.	*Microtendipes pedellus* group
63	Arthropoda	Insecta	Diptera	Nematocera	Chironomidae	Chironominae	Chironomini	*Parachironomus* sp.	
64	Arthropoda	Insecta	Diptera	Nematocera	Chironomidae	Chironominae	Chironomini	*Polypedilum* sp.	
65	Arthropoda	Insecta	Diptera	Nematocera	Chironomidae	Chironominae	Chironomini	*Sergentia* sp.	

Table A1. Benthic macroinvertebrate community detected at stream sites along McCarthy Road, Wrangell-St. Elias National Park and Preserve, Alaska, 2007–08.—Continued

Count	Phylum	Class	Order	SubOrder	Family	SubFamily	Tribe	Genus	Species
66	Arthropoda	Insecta	Diptera	Nematocera	Chironomidae	Chironominae	Chironomini	Stictochironomus sp.	
67	Arthropoda	Insecta	Diptera	Nematocera	Chironomidae	Chironominae	Tanytarsini	Constempellina sp.	
68	Arthropoda	Insecta	Diptera	Nematocera	Chironomidae	Chironominae	Tanytarsini	Micropsectra sp.	
69	Arthropoda	Insecta	Diptera	Nematocera	Chironomidae	Chironominae	Tanytarsini	Rheotanytarsus sp.	
70	Arthropoda	Insecta	Diptera	Nematocera	Chironomidae	Chironominae	Tanytarsini	Stempellinella sp.	
71	Arthropoda	Insecta	Diptera	Nematocera	Chironomidae	Diamesinae	Diamesini	Diamesa sp.	
72	Arthropoda	Insecta	Diptera	Nematocera	Chironomidae	Diamesinae	Diamesini	Pagastia sp.	
73	Arthropoda	Insecta	Diptera	Nematocera	Chironomidae	Diamesinae	Diamesini	Potthastia sp.	Potthastia longimana group
74	Arthropoda	Insecta	Diptera	Nematocera	Chironomidae	Diamesinae	Diamesini	Pseudodiamesa sp.	
75	Arthropoda	Insecta	Diptera	Nematocera	Chironomidae	Orthocladiinae		Brillia sp.	
76	Arthropoda	Insecta	Diptera	Nematocera	Chironomidae	Orthocladiinae		Cardiocladius sp.	
77	Arthropoda	Insecta	Diptera	Nematocera	Chironomidae	Orthocladiinae		Chaetocladius sp.	
78	Arthropoda	Insecta	Diptera	Nematocera	Chironomidae	Orthocladiinae		Corynoneura sp.	
79	Arthropoda	Insecta	Diptera	Nematocera	Chironomidae	Orthocladiinae		Cricotopus sp.	Cricotopus bicinctus group
80	Arthropoda	Insecta	Diptera	Nematocera	Chironomidae	Orthocladiinae		Diplocladius sp.	
81	Arthropoda	Insecta	Diptera	Nematocera	Chironomidae	Orthocladiinae		Eukiefferiella sp.	Eukiefferiella brevicalcar
82	Arthropoda	Insecta	Diptera	Nematocera	Chironomidae	Orthocladiinae		Eukiefferiella sp.	Eukiefferiella claripennis group sp. C
83	Arthropoda	Insecta	Diptera	Nematocera	Chironomidae	Orthocladiinae		Eukiefferiella sp.	Eukiefferiella coerulescens group
84	Arthropoda	Insecta	Diptera	Nematocera	Chironomidae	Orthocladiinae		Eukiefferiella sp.	Eukiefferiella devonica group
85	Arthropoda	Insecta	Diptera	Nematocera	Chironomidae	Orthocladiinae		Eukiefferiella sp.	Eukiefferiella gracei group
86	Arthropoda	Insecta	Diptera	Nematocera	Chironomidae	Orthocladiinae		Eukiefferiella sp.	Eukiefferiella tirolensis
87	Arthropoda	Insecta	Diptera	Nematocera	Chironomidae	Orthocladiinae		Heterotanytarsus sp.	
88	Arthropoda	Insecta	Diptera	Nematocera	Chironomidae	Orthocladiinae		Heterotrissocladius sp.	Heterotrissocladius marcidus group
89	Arthropoda	Insecta	Diptera	Nematocera	Chironomidae	Orthocladiinae		Hydrobaenus sp.	
90	Arthropoda	Insecta	Diptera	Nematocera	Chironomidae	Orthocladiinae		Limnophyes sp.	
91	Arthropoda	Insecta	Diptera	Nematocera	Chironomidae	Orthocladiinae		Metriocnemus sp.	
92	Arthropoda	Insecta	Diptera	Nematocera	Chironomidae	Orthocladiinae		Orthocladius (Euorthocladius) sp.	

Table A1. Benthic macroinvertebrate community detected at stream sites along McCarthy Road, Wrangell-St. Elias National Park and Preserve, Alaska, 2007–08.—Continued

Count	Phylum	Class	Order	SubOrder	Family	SubFamily	Tribe	Genus	Species
93	Arthropoda	Insecta	Diptera	Nematocera	Chironomidae	Orthocladiinae		*Orthocladius Complex* sp.	
94	Arthropoda	Insecta	Diptera	Nematocera	Chironomidae	Orthocladiinae		*Orthocladius* sp.	
95	Arthropoda	Insecta	Diptera	Nematocera	Chironomidae	Orthocladiinae		*Orthocladius* sp.	*Orthocladius (Euortho.) rivicola*
96	Arthropoda	Insecta	Diptera	Nematocera	Chironomidae	Orthocladiinae		*Parakiefferiella* sp.	
97	Arthropoda	Insecta	Diptera	Nematocera	Chironomidae	Orthocladiinae		*Parametriocnemus* sp.	
98	Arthropoda	Insecta	Diptera	Nematocera	Chironomidae	Orthocladiinae		*Paraphaenocladius* sp.	
99	Arthropoda	Insecta	Diptera	Nematocera	Chironomidae	Orthocladiinae		*Psectrocladius* sp.	
100	Arthropoda	Insecta	Diptera	Nematocera	Chironomidae	Orthocladiinae		*Rheocricotopus* sp.	
101	Arthropoda	Insecta	Diptera	Nematocera	Chironomidae	Orthocladiinae		*Synorthocladius* sp.	
102	Arthropoda	Insecta	Diptera	Nematocera	Chironomidae	Orthocladiinae		*Thienemanniella* sp.	
103	Arthropoda	Insecta	Diptera	Nematocera	Chironomidae	Orthocladiinae		*Tokunagaia* sp.	
104	Arthropoda	Insecta	Diptera	Nematocera	Chironomidae	Orthocladiinae		*Tvetenia* sp.	*Tvetenia bavarica*
105	Arthropoda	Insecta	Diptera	Nematocera	Chironomidae	Podonominae	Boreochlini	*Boreochlus* sp.	
106	Arthropoda	Insecta	Diptera	Nematocera	Chironomidae	Podonominae	Boreochlini	*Trichotanypus* sp.	
107	Arthropoda	Insecta	Diptera	Nematocera	Chironomidae	Prodiamesinae		*Prodiamesa* sp.	
108	Arthropoda	Insecta	Diptera	Nematocera	Chironomidae	Tanypodinae	Pentaneurini	*Paramerina* sp.	
109	Arthropoda	Insecta	Diptera	Nematocera	Chironomidae	Tanypodinae	Pentaneurini	*Thienemannimyia gr* sp.	
110	Arthropoda	Insecta	Diptera	Nematocera	Chironomidae	Tanypodinae	Procladini	*Procladius* sp.	
111	Arthropoda	Insecta	Diptera	Nematocera	Dixidae			*Dixella* sp.	
112	Arthropoda	Insecta	Diptera	Nematocera	Psychodidae	Psychodinae		*Pericoma/ Telmatoscopus* sp.	
113	Arthropoda	Insecta	Diptera	Nematocera	Simuliidae	Simuliinae		*Prosimulium* sp.	
114	Arthropoda	Insecta	Diptera	Nematocera	Simuliidae	Simuliinae		*Simulium* sp.	
115	Arthropoda	Insecta	Diptera	Nematocera	Simuliidae	Simuliinae	Prosimuliini	*Greniera* sp.	
116	Arthropoda	Insecta	Diptera	Nematocera	Tipulidae	Limoniinae		*Dicranota* sp.	
117	Arthropoda	Insecta	Diptera	Nematocera	Tipulidae	Limoniinae	Eriopterini	*Rhabdomastix* sp.	*Rhabdomastix setigera*
118	Arthropoda	Insecta	Diptera	Nematocera	Tipulidae	Limoniinae / Limoniinae	Eriopterini	*Rhabdomastix* sp.	*Rhabdomastix tricophora*
119	Arthropoda	Insecta	Diptera	Brachycera	Empididae	Clinocerinae		*Clinocera* sp.	
120	Arthropoda	Insecta	Diptera	Brachycera	Empididae	Clinocerinae		*Trichoclinocera* sp.	
121	Arthropoda	Insecta	Diptera	Brachycera	Empididae	Empidinae		*Oreogeton* sp.	
122	Arthropoda	Insecta	Diptera	Brachycera	Empididae	Empidinae		*Chelifera/Metachela* sp.	
123	Arthropoda	Insecta	Diptera	Brachycera	Empididae	Hemerodrominae			
124	Arthropoda	Insecta	Diptera	Brachycera	Sciomyzidae				
125	Arthropoda	Insecta	Diptera	Brachycera	Syrphidae				

Appendix B. Periphytic Algae Taxa Present in Qualitative Multi-Habitat Samples Collected at Stream Sites along McCarthy Road, Wrangell-St. Elias National Park and Preserve, Alaska, 2007–08

Table B1.   Periphytic algae taxa present in Qualitative Multi-Habitat samples collected at stream sites along McCarthy Road, Wrangell-St. Elias National Park and Preserve, Alaska.

Algae group	Phylum	Class	Family	Genus species
Lakina River (10-02-07)				
Diatoms	Chrysophyta	Bacillariophyceae	Achnanthaceae	*Cocconeis placentula*
Diatoms	Chrysophyta	Bacillariophyceae	Achnanthaceae	*Cocconeis pseudolineata*
Diatoms	Chrysophyta	Bacillariophyceae	Achnanthidiaceae	*Achnanthidium minutissimum*
Diatoms	Chrysophyta	Bacillariophyceae	Bacillariaceae	*Nitzschia paleacea*
Diatoms	Chrysophyta	Bacillariophyceae	Catenulaceae	*Amphora pediculus*
Diatoms	Chrysophyta	Bacillariophyceae	Cymbellaceae	*Encyonema minutum*
Diatoms	Chrysophyta	Bacillariophyceae	Cymbellaceae	*Encyonema silesiacum*
Diatoms	Chrysophyta	Bacillariophyceae	Fragilariaceae	*Diatoma mesodon*
Diatoms	Chrysophyta	Bacillariophyceae	Fragilariaceae	*Diatoma moniliformis*
Diatoms	Chrysophyta	Bacillariophyceae	Fragilariaceae	*Fragilaria vaucheriae*
Diatoms	Chrysophyta	Bacillariophyceae	Fragilariaceae	*Hannaea arcus*
Diatoms	Chrysophyta	Bacillariophyceae	Fragilariaceae	*Meridion circulare*
Diatoms	Chrysophyta	Bacillariophyceae	Fragilariaceae	*Staurosira construens*
Diatoms	Chrysophyta	Bacillariophyceae	Fragilariaceae	*Staurosirella pinnata*
Diatoms	Chrysophyta	Bacillariophyceae	Fragilariaceae	*Synedra ulna*
Diatoms	Chrysophyta	Bacillariophyceae	Gomphonemataceae	*Gomphonema drutelingense*
Diatoms	Chrysophyta	Bacillariophyceae	Gomphonemataceae	*Gomphonema longilineare*
Diatoms	Chrysophyta	Bacillariophyceae	Gomphonemataceae	*Reimeria sinuata*
Diatoms	Chrysophyta	Bacillariophyceae	Naviculaceae	*Navicula cryptotenella*
Yellow-Green	Chrysophyta	Chrysophyceae	Hydruraceae	*Hydrurus foetidus*
Red Algae	Rhodophyta	(Undetermined)		Unknown Rhodophyte
Long Lake Inflow-Tributary 1 (08-06-08)				
Diatoms	Chrysophyta	Bacillariophyceae	Achnanthaceae	*Psammothidium chlidanos*
Diatoms	Chrysophyta	Bacillariophyceae	Achnanthidiaceae	*Achnanthidium deflexum*
Diatoms	Chrysophyta	Bacillariophyceae	Achnanthidiaceae	*Achnanthidium minutissimum*
Diatoms	Chrysophyta	Bacillariophyceae	Achnanthidiaceae	*Achnanthidium pyrenaicum*
Diatoms	Chrysophyta	Bacillariophyceae	Achnanthidiaceae	*Eucocconeis laevis*
Diatoms	Chrysophyta	Bacillariophyceae	Catenulaceae	*Amphora inariensis*
Diatoms	Chrysophyta	Bacillariophyceae	Catenulaceae	*Amphora pediculus*
Diatoms	Chrysophyta	Bacillariophyceae	Cymbellaceae	*Cymbella cymbiformis*
Diatoms	Chrysophyta	Bacillariophyceae	Cymbellaceae	*Cymbella tumida*
Diatoms	Chrysophyta	Bacillariophyceae	Fragilariaceae	*Meridion circulare*
Diatoms	Chrysophyta	Bacillariophyceae	Fragilariaceae	*Synedra acus*
Diatoms	Chrysophyta	Bacillariophyceae	Fragilariaceae	*Synedra ulna*
Diatoms	Chrysophyta	Bacillariophyceae	Gomphonemataceae	*Gomphonema drutelingense*
Diatoms	Chrysophyta	Bacillariophyceae	Gomphonemataceae	*Reimeria sinuata*
Diatoms	Chrysophyta	Bacillariophyceae	Naviculaceae	*Navicula* sp.
Diatoms	Chrysophyta	Bacillariophyceae	Naviculaceae	*Navicula minima*
Diatoms	Chrysophyta	Bacillariophyceae	Neidiaceae	*Neidium hercynicum*
Diatoms	Chrysophyta	Bacillariophyceae	Pinnulariaceae	*Caloneis bacillum*
Diatoms	Chrysophyta	Bacillariophyceae	Sellaphoraceae	*Sellaphora pupula*

Table B1. Periphytic algae taxa present in Qualitative Multi-Habitat samples collected at stream sites along McCarthy Road, Wrangell-St. Elias National Park and Preserve, Alaska.—Continued

Algae group	Phylum	Class	Family	Genus species
			Long Lake Inflow-Tributary 2 (07-30-07)	
Green Algae	Chlorophyta	Chlorophyceae	Chaetophoraceae	*Stigeoclonium* sp.
Diatoms	Chrysophyta	Bacillariophyceae	Achnanthaceae	*Cocconeis pediculus*
Diatoms	Chrysophyta	Bacillariophyceae	Achnanthaceae	*Cocconeis placentula*
Diatoms	Chrysophyta	Bacillariophyceae	Achnanthaceae	*Cocconeis pseudolineata*
Diatoms	Chrysophyta	Bacillariophyceae	Achnanthaceae	*Karayevia clevei*
Diatoms	Chrysophyta	Bacillariophyceae	Achnanthaceae	*Platessa conspicua*
Diatoms	Chrysophyta	Bacillariophyceae	Achnanthidiaceae	*Achnanthidium minutissimum*
Diatoms	Chrysophyta	Bacillariophyceae	Achnanthidiaceae	*Achnanthidium rivulare*
Diatoms	Chrysophyta	Bacillariophyceae	Bacillariaceae	*Nitzschia angustata*
Diatoms	Chrysophyta	Bacillariophyceae	Bacillariaceae	*Nitzschia dissipata*
Diatoms	Chrysophyta	Bacillariophyceae	Bacillariaceae	*Nitzschia linearis*
Diatoms	Chrysophyta	Bacillariophyceae	Bacillariaceae	*Nitzschia palea*
Diatoms	Chrysophyta	Bacillariophyceae	Bacillariaceae	*Nitzschia pumila*
Diatoms	Chrysophyta	Bacillariophyceae	Bacillariaceae	*Nitzschia sociabilis*
Diatoms	Chrysophyta	Bacillariophyceae	Catenulaceae	*Amphora pediculus*
Diatoms	Chrysophyta	Bacillariophyceae	Cymbellaceae	*Cymbella* sp.
Diatoms	Chrysophyta	Bacillariophyceae	Cymbellaceae	*Cymbella cymbiformis*
Diatoms	Chrysophyta	Bacillariophyceae	Fragilariaceae	*Diatoma mesodon*
Diatoms	Chrysophyta	Bacillariophyceae	Fragilariaceae	*Fragilaria tenera*
Diatoms	Chrysophyta	Bacillariophyceae	Fragilariaceae	*Fragilaria vaucheriae*
Diatoms	Chrysophyta	Bacillariophyceae	Fragilariaceae	*Meridion circulare*
Diatoms	Chrysophyta	Bacillariophyceae	Fragilariaceae	*Staurosirella leptostauron*
Diatoms	Chrysophyta	Bacillariophyceae	Fragilariaceae	*Staurosirella pinnata*
Diatoms	Chrysophyta	Bacillariophyceae	Fragilariaceae	*Synedra ulna*
Diatoms	Chrysophyta	Bacillariophyceae	Gomphonemataceae	*Gomphonema drutelingense*
Diatoms	Chrysophyta	Bacillariophyceae	Naviculaceae	*Geissleria acceptata*
Diatoms	Chrysophyta	Bacillariophyceae	Naviculaceae	*Navicula antonii*
Diatoms	Chrysophyta	Bacillariophyceae	Naviculaceae	*Navicula cryptocephala*
Diatoms	Chrysophyta	Bacillariophyceae	Naviculaceae	*Navicula cryptotenella*
Diatoms	Chrysophyta	Bacillariophyceae	Naviculaceae	*Navicula lanceolata*
Diatoms	Chrysophyta	Bacillariophyceae	Rhopalodiaceae	*Epithemia adnata*
Diatoms	Chrysophyta	Bacillariophyceae	Rhopalodiaceae	*Rhopalodia gibba*
Diatoms	Chrysophyta	Bacillariophyceae	Stauroneidaceae	*Stauroneis smithii*
Diatoms	Chrysophyta	Bacillariophyceae	Stephanodiscaceae	*Cyclotella rossii*
Diatoms	Chrysophyta	Bacillariophyceae	Thalassiosiraceae	*Puncticulata radiosa*
Blue-Green	Cyanophyta	Myxophyceae	Chroococcaceae	*Chroococcus* sp.
Blue-Green	Cyanophyta	Myxophyceae	Phormidiaceae	*Phormidium* sp.
Blue-Green	Cyanophyta	Myxophyceae	Pseudanabaenaceae	*Pseudanabaena* sp.
Blue-Green	Cyanophyta	Myxophyceae	Rivulariaceae	*Calothrix* sp.
Red Algae	Rhodophyta	Rhodophyceae	Batrachospermaceae	*Batrachospermum* sp.
			Long Lake Outflow (07-31-07)	
Green Algae	Chlorophyta	(undetermined)		Unknown Chlorophyte
Green Algae	Chlorophyta	Chlorophyceae	Chaetophoraceae	*Stigeoclonium* sp.
Green Algae	Chlorophyta	Chlorophyceae	Cladophoraceae	*Cladophora glomerata*
Green Algae	Chlorophyta	Chlorophyceae	Cylindrocapsaceae	*Cylindrocapsa geminella*
Green Algae	Chlorophyta	Chlorophyceae	Desmidiaceae	*Closterium moniliferum*
Green Algae	Chlorophyta	Chlorophyceae	Desmidiaceae	*Staurastrum* sp.
Green Algae	Chlorophyta	Chlorophyceae	Oedogoniaceae	*Oedogonium* sp.
Green Algae	Chlorophyta	Chlorophyceae	Oocystaceae	*Ankistrodesmus fusiformis*
Green Algae	Chlorophyta	Chlorophyceae	Tetrasporaceae	*Tetraspora gelatinosa*
Green Algae	Chlorophyta	Chlorophyceae	Zygnemataceae	*Mougeotia* sp.
Green Algae	Chlorophyta	Chlorophyceae	Zygnemataceae	*Spirogyra* sp.
Diatoms	Chrysophyta	Bacillariophyceae	Achnanthaceae	*Cocconeis pediculus*
Diatoms	Chrysophyta	Bacillariophyceae	Achnanthaceae	*Cocconeis placentula*